Assessing and Managing Groundwater in Different Environments

T0144234

Selected papers on hydrogeology

19

Series Editor: Dr. Nick S. Robins
Editor-in-Chief IAH Book Series, British Geological Survey, Wallingford, UK

INTERNATIONAL ASSOCIATION OF HYDROGEOLOGISTS

Assessing and Managing Groundwater in Different Environments

Editors

Jude Cobbing
SLR Consulting and Nelson Mandela Metropolitan University, South Africa

Shafick Adams
Water Research Commission, Pretoria, South Africa

Ingrid Dennis
North West University, Potchestroom, South Africa

Kornelius Riemann
Umvoto Consulting, Cape Town, South Africa

CRC Press
Taylor & Francis Group
Boca Raton London New York

CRC Press is an imprint of the
Taylor & Francis Group, an **informa** business

A BALKEMA BOOK

**GROUND WATER
DIVISION**
www.gwd.org.za

The editors would like to thank the Groundwater Division of the Geological Society of South Africa, who assisted in all aspects of publication and work closely with the International Association of Hydrogeologists.

CRC Press
Taylor & Francis Group
6000 Broken Sound Parkway NW, Suite 300
Boca Raton, FL 33487-2742

First issued in paperback 2019

© 2013 by Taylor & Francis Group, LLC
CRC Press is an imprint of Taylor & Francis Group, an Informa business

No claim to original U.S. Government works

ISBN-13: 978-1-138-00100-8 (hbk)
ISBN-13: 978-0-367-37940-7 (pbk)

Library of Congress Cataloging-in-Publication Data
Applied for

**Visit the Taylor & Francis Web site at
http://www.taylorandfrancis.com**

**and the CRC Press Web site at
http://www.crcpress.com**

Table of contents

About the editors

Mr **Jude Cobbing** is currently working on a PhD in rural groundwater supply sustainability at Nelson Mandela Metropolitan University in Port Elizabeth, South Africa. A hydrogeologist at the British Geological Survey from 2000 to 2005, Jude has worked as a researcher and consultant in South Africa since then. His interests include rural water supply hydrogeology in Africa, groundwater management, groundwater data, and the training of hydrogeologists. He has worked in Botswana, South Africa, Malawi, Madagascar, Nigeria, Ghana, Tanzania, Bangladesh, the Democratic Republic of the Congo and the United Kingdom. He has a BSc in Geology from the University of Cape Town, a PGCE from the University of South Africa, and an MSc in hydrogeology from London University.

Dr **Shafick Adams** is currently employed by the Water Research Commission in Pretoria, South Africa where he manages research projects related to groundwater and water resources protection. He holds a PhD and MSc (*cum laude*) from the University of the Western Cape where he lectured for a few years prior to his appointment to the Water Research Commission. He currently chairs the Groundwater Division of the Geological Society of South Africa and is co-chair of the International Water Association's Groundwater Restoration and Management Specialist Group. Some of his research interests are in groundwater recharge assessments, chemical characterisation of groundwater, capacity development, and water resource management.

Prof. **Ingrid Dennis** started her career as a hydrogeologist at the CSIR in Stellenbosch, South Africa. She later moved to the Institute for Groundwater Studies at the University of the Free State to complete her PhD in hydrogeology and work as a lecturer and researcher. She was appointed as the Director of the Institute for Groundwater Studies in 2007. Prof Dennis joined the North-West University in July 2011 as an Associate Professor and heads up the Centre for Water Sciences and Management.

Prof. **Dennis** has over 15 years of experience as both a consultant and a researcher. She specialises in analytical and numerical modelling, hydrogeological baseline studies, mining and industrial impact assessments. She presents numerous training courses throughout South Africa (to government officials and consultants) concerning the management of groundwater resources and impacts on groundwater systems.

 Born in Hannover, Germany, **Kornelius Riemann** obtained his MSc in Hydrogeology from the University of Kiel. He joined Umvoto Africa in 2002 following completion of his PhD at the University of the Free State. His PhD research in South Africa dealt with new approaches and methods for estimating aquifer parameters in fractured rock, based on fractal analysis of test pumping and tracer-test data. Dr Riemann has extended experience with work in groundwater assessment studies, water resource evaluations, hydrochemistry, hydrogeological interpretation, and contamination investigation. He has led several studies with respect to surface water – groundwater interaction and the development of monitoring networks, and is responsible for developing and implementing strategies for adaptive groundwater management, improved resource evaluation, sophisticated monitoring of response to abstraction, water quality and ecological impacts, and modelling for sustainable resource management.

Introduction/setting the scene: 2011 GWD/IAH book intro

J. Cobbing, S. Adams, I. Dennis & K. Riemann

This book is a collection of Selected Papers presented at the Biennial Conference of the Ground Water Division (GWD) of the Geological Society of South Africa, held in Pretoria in October 2011. Looking back over past IAH Congress publications and similar collections one is struck by how many times the 'hidden resource' of groundwater is closely linked to other themes or fields of study: Climate, human development, land-use management, the urban environment, ecosystems – all are associated with groundwater. This is no accident, since groundwater is integral to many, many human and environmental systems. Indeed, there appears to be a growing realisation that some of the most pressing physical problems in the field of hydrogeology such as over-abstraction, salinization or pollution can only really be solved by taking a multi-disciplinary approach to the issues. Whilst a 'technical' solution may be deciphered, the larger challenge usually lies in the sustainably-funded and widely-accepted implementation of that measure. In this sense today's hydrogeologist may need to have at least some practical knowledge of environmental science, economics, management theory, mining, environmental law, and water treatment – amongst other disciplines.

Different countries may have subtly different approaches and expectations when it comes to managing groundwater. A delegation of Sudanese visitors on a tour of South Africa's drought-struck southern Cape region in 2010 politely coined the phrase 'green drought' to refer to what was – to them – a distinctly underwhelming crisis. A distinguished Australian professor at a meeting in South Africa perplexed her audience by asserting that peer pressure and the fear of being 'dobbed-in' by one's neighbours encouraged Australian borehole owners to report accurate abstraction quantities to local authorities. And French and British hydrogeologists have yet to agree on a common nomenclature for the most important water bearing units of the prolific Cretaceous Chalk aquifer found in both countries. One of the great strengths of the GWD Conference was the range of nationalities, specialities and experiences of the presenters and attendees. From radionuclide transport in German groundwater to the sustainability of the African basement aquifer or the management of groundwater resources for emergency purposes in Slovakia, the conference was an opportunity to share experiences and to consider hybrid solutions in a world that increasingly confounds the old 'developed' versus 'developing' tags. There is great value in such sharing and collaboration: after all, as the cockerel said to the hen on seeing an ostrich egg: 'I am not criticising; I am not belittling: I am merely pointing out what is being done elsewhere!'

Some samples of the papers presented showcase the range of topics:

The first paper in this collection, by conference keynote speaker Dr Andrew Stone of the American Groundwater Trust, confirms the close connection of groundwater to wider hydrological, environmental, planning and economic issues. Around half of all drinking water in the USA is derived from groundwater, and challenges range from more established ones such as controlling leakages, setting appropriate pricing and tightening quality regulations, to newer issues related to climate change and hydraulic fracturing for shale gas. As elsewhere, there is a continued drift from supply-side solutions (i.e. providing more water) to using what we have more effectively – leading to advances in aquifer storage and recovery, desalination technologies, reclamation of wastewater, and public education. All call for more comprehensive and effective planning, and a 'holistic' view. Dr Danie Vermeulen provides a South African perspective on the shale gas fracking issue, currently a controversial topic in that country, in his paper 'Preliminary assessment of water-supply availability with regard to potential shale-gas development in the Karoo region of South Africa'. In her paper 'The challenges of municipal groundwater supply from private land' Helen Seyler describes the practical issues and trade-offs inherent in developing groundwater for public supply, under the provisions of the South African National Water Act. Roger Parsons' paper 'Development of emergency water supplies for the drought-impacted southern Cape coastal region of South Africa - observations while abstracting saline water for desalination' adds another dimension to the study of groundwater for municipal supply.

In their paper 'Hydrogeological study for sustainable water-resource exploitation – Ibo Island, Mozambique' Vilanova *et al.* report on the groundwater supply of a small African island, and make policy recommendations. These recommendations are founded on a thorough understanding of the hydrogeology but will require the participation of the wider society and of other professionals if they are to be sustainably implemented.

Several papers deal with the application of numerical methods for 'real world' problems. Seyler and Hay discuss the use of models in the Western Cape to 'break the logjam' of uncertainty and assist in the rollout of municipal groundwater supply schemes. Witthueser and Konig describe the use of stochastic fracture generation in modelling underground mines. Fluegge *et al.* describe work on the implications of changing climate (e.g. sea-level rises, permafrost formation) on our assumptions regarding radionuclide transport in underground waste repositories.

As the outcome of a groundwater conference with a wide mandate, this eclectic collection of papers is perhaps a snapshot of modern hydrogeology – a profession in which highly technical methods and approaches sit side-by-side with overlapping (and sometimes complex and even intractable) legal, social, institutional and governance considerations.

Chapter 1

Water-resource issues in the United States and the changing focus on groundwater

Andrew Stone
American Ground Water Trust, Concord, New Hampshire, USA

ABSTRACT

In America, as in most countries in the world, the challenges of supply and demand for water keep a strong focus on the critical role of groundwater-resource management. Strategies to conserve, recharge and structure withdrawals have not only become more sophisticated but are increasingly connected to wider hydrological and environmental issues. Regulatory measures for drinking-water supply and for environmental restoration impose new challenges for water managers and their hydrogeological advisors. The causes (drivers) of trends in American hydrogeology include factors such as water scarcity, climate uncertainty, energy costs, carbon footprint concerns, ecological requirements and new concerns about contaminant concentrations greater than human-health thresholds.

1.1 INTRODUCTION

Groundwater is a vital economic and environmental asset in the US. Fifty three per cent of the nation's drinking water is derived from groundwater sources, with 140 000 public water systems across the United States served by groundwater from public wells (Toccalino *et al.*, 2010) and approximately 40 million citizens supplied from private wells. In addition to groundwater use by industry and power generators, much of the US agricultural economy is dependent on groundwater for irrigation. Traditional roles for groundwater professionals have been to identify groundwater resources, to assess their potential for sustainable groundwater development, to characterise instances of contamination, and to develop remediation strategies. The roles of 21st century groundwater professionals are more closely bound to holistic resource management issues than in decades past. There are new and emerging 'drivers' of water-resource issues. Groundwater specialists are increasingly involved in working with other professions in the application of their skills to problem solving.

According to a recent report (Johnson Foundation, 2010) the US is facing major water issues. The report was produced by a diverse group of water users and policy specialists and presented in September 2010 to the Obama Administration at a meeting of federal agencies convened by the White House Council on Environmental Policy. 'There was broad consensus that our current path will, unless changed, lead us to a national freshwater crisis. This reality encompasses a wide array of challenges ... that collectively amount to a tenuous trajectory for the future of the nation's freshwater resources.' (CEQ, 2009).

As water-resource decisions become increasingly bound together with economic, energy, ecological and planning issues, groundwater professionals need to ensure that any policy involving or impacting water resources is based on good science. For example, as the US seeks greater independence from energy imports, the water/energy nexus arises over arguments about water needs for biofuel crops, water needs for hydraulic fracturing in the development of shale gas and coal-bed methane and the associated environmental challenges to water supply. The growing trend of geothermal heating and cooling installations in North America has provided a huge boost to the well-construction industry and is a developing field for hydrogeologists in working with engineers to characterise subsurface heat-transfer capabilities. Recent health studies that indicate a need for stricter water quality standards for contaminants such as arsenic, perchlorate and hexavalent chromium (chromium VI) will potentially reduce the inventory of usable aquifers and hence pose new treatment and management challenges.

Aquifer recharge is an arena of groundwater technology that is likely to emerge as a solution to some of the water storage and management challenges. Water reuse from wastewater treatment and stormwater capture for recharge are the only available sources in some areas to meet supply demand, prompting needs for predictions of hydrological and hydrochemical impacts. As with many other technical issues, the 'education' of citizens, communities and decision-makers is a prerequisite for streamlining the regulatory process for project implementation and for giving private or public investors confidence in financing aquifer-recharge projects.

1.2   WATER USE, WATER INDUSTRY, PRICING AND INFRASTRUCTURE

Water utilities in the US do an excellent job in delivering fresh, clean drinking water at a minimal cost. However, one of the major challenges for water supply is the decaying infrastructure, with an average of 17% of water treated to drinking standards being 'lost' from the distribution system; although an unintended consequence of the loss may be a gain for groundwater recharge. It is estimated that 26.5×10^6 m^3 (seven billion gallons) each day is lost through leakage. This water is sourced, treated, stored and distributed, then wasted (Maxwell, 2011). Many urban and suburban areas in the US were developed in the 19th and early 20th century and maintenance and replacement have not kept pace with infrastructure decay. Various government agencies and water-industry organisations show 20-year infrastructure needs in the US of between 700 billion and one trillion dollars. There is little argument about the huge capital investment needs, but there is considerable uncertainty about potential sources of the capital required. The costs of treating and pumping water that does not generate revenue and the need to maintain production capacity greater than sales, contribute to the disparity between replacement needs and available financial resources. This would not be so much of a problem if water was realistically priced. In the US, as in other countries, water is much more valuable than its price. Water pricing is rarely related to its true cost which should include all aspects of development, operation and maintenance. US families typically use 454 m^3 (120 000 gallons) a year for a cost of around US\$350 (US\$1/d) (Maxwell, 2011).

In addition to a slow but steady appreciation of the value of water conservation, there is also an increasing public awareness in the US of water 'footprints' that incorporate the imbedded water in products, especially food products. However, water conservation is not a permanent solution. Realistic end-user pricing is needed to generate the revenue for public sector or private sector water utility investment in developing supply, treatment and storage infrastructure. A major beneficiary from investment capital would be groundwater-storage projects that are likely to be far more cost-effective than surface-water engineering solutions. The author of a recent publication on water issues makes the general case on pricing, '*If you had to pick one thing to fix about water, one thing that would help you fix everything else – scarcity, unequal distribution, misuse, waste, skewed priorities, resistance to reuse, shortsighted exploitation of natural water resources – that one thing is price. The right price changes how we see everything else about water.*' (Fishman, 2011).

1.3   CLIMATE-CHANGE IMPACTS

Now that the reality of climate change is seen as 'unequivocal' (Bates *et al.*, 2008), even the most doubting water managers are incorporating greater hydrological uncertainty into planning. Sea-level changes will impact the freshwater geometry of coastal aquifers. Changes in intensity, duration, frequency and type of meteorological input will impact recharge potential to aquifers. Population and agricultural production shifts will change demand patterns. Temperature changes may create algal and eutrophication responses that increase surface-water source costs. Less certainty in predicting climate will constrain overall water planning supply and demand estimates. Hydrogeologists will find an increasingly important role in the modelling of relationships among hydrological, geological, infrastructure, regulatory, environmental and socio-economic factors in water planning. Accurate characterisation of the groundwater-storage capacities and recharge potential of aquifers of all sizes is likely to become more important as aquifers become critical water-management buffers in the likely game-changing impacts of increasing climatic variability.

1.4   DEMAND PREDICTIONS – CONSERVATION

With the US population growing by 7 000 people each day, and the increasing variability of hydrological input associated with climate change, many US water managers are exploring options that involve groundwater as part of the storage and delivery mix. Desalination of ocean water or brackish groundwater, especially with greater efficiencies in desalination technology, is emerging as a potential source of recharge water for aquifer storage. The reuse and reclamation of wastewater by either direct or indirect reuse is viewed by some as the only dependable source for 'new' supply, with recharge in aquifers serving as the 'indirect reuse' buffer. Green infrastructure in urban and suburban areas, which is designed to slow runoff and delay pressure on stormwater facilities, is prompting a form of multi-point on-site recharge by reducing overall urban impermeability. Green building water harvesting with on-site storage of rain water can reduce household demand and help reduce peak storm runoff.

Water conservation is the easy-to-implement 'low-hanging fruit' to delay water-utility investment in delivery infrastructure. Influencing behaviour change by education, and by offering inducements such as paying for homeowners to give up lawn grass, and by tariff penalties for excess water use have had a dramatic impact in the arid southwest. For example, Denver, Las Vegas and Albuquerque have developed effective education and incentive programmes that have reduced per capita consumption. However, there are very few areas where there is major water need that can 'save their way to prosperity.' Water managers also have to resolve the conundrum of lower use, lower revenues and therefore less capital available for infrastructure improvements.

1.5   WATER-RESOURCE PLANNING

The *ad hoc* approach to anticipating and meeting water-supply demands is rapidly being replaced by comprehensive approaches and sophisticated techniques. One approach developed by the American Water Works Association is integrated resource planning (IRP). This holistic approach to water supply involves least-cost analysis of demand-side and supply-side management options and also considers socioeconomic and environmental issues. The process involves assessment of management options to achieve cost-effective supply objectives. Details of the process have been developed and published by the American Water Works Association (Maddaus, 2008). In a very simplified form, three stages of the process can be identified with each stage presenting opportunities for hydrogeological expertise. For any given region or supply area, data for population trends, economic forecasts and meteorological input projections are a prerequisite for the IRP planning process. At all stages of the IRP process there is opportunity for public input.

Forecast demand
This stage requires supply-side quantification of the safe-yield of current water sources and an assessment of the ability of conservation measures and water reclamation to reduce the demand-side.

Develop resource strategies
This stage takes current supply-side reliability assessments and combines them with the hydrological and economic feasibility of new and imported water and the limitation thresholds on any additional measures to reduce demand, such as education, incentives or pricing.

Evaluate and rank options
Once all data sets and forecasts are obtained, the heart of the IRP process involves systematic assessment iterations of engineering feasibility, cost, financing availability, water quality and environmental impacts and public acceptance.

The identification and assessment of new supply-side resources and calculations of the optimised potential from existing well fields is a clear role for groundwater professionals in IRP. Conjunctive use of surface-water and groundwater sources and the growing emergence of 'recycled' and imported sources for recharge will likely be of growing importance in providing solutions. To solve many water shortages there will be exchange and sale of water rights among land-owners and utilities, principally involving the transfer from agricultural to urban use. These transfers will involve aggregated transactions worth millions of dollars. With water and money at stake, hydrogeological models will be a key component of accurate aquifer quantification and 'molecule tracking' of recharged sources.

1.6 ROLE OF PRIVATE SECTOR

One element of the solution to the US infrastructure investment needed to maintain current supply and develop new resources is to allow for a larger role of the private sector. Currently, approximately 80 million Americans are directly supplied independently or via private companies. Approximately 12% to 15% of the US population receives water from private companies (Maxwell, 2011) and an additional 13% of the population is self-supplied from on-site home wells with the cost borne by the homeowner and the value of the well system bound into the equity value of the home. The homeowner has responsibility for all operation and maintenance costs. However, in the US, as in many other parts of the world, there is public resistance to for-profit companies having anything to do with water delivery. The philosophical basis for such objections seems to be based on the belief that water is a basic human right, and as such, should be off-limits to the private sector. Water is certainly a basic human need and governments (at all levels of jurisdiction) have a prime responsibility to ensure that their citizens have affordable access to water. Whether or not water is a 'right' is probably a semantic argument (what about food, clothing, shelter, broadband access?). However, and it is a big however, ensuring that citizens have affordable access should not rule out the potential involvement of the private sector in investing in infrastructure or as licensed managers of source, storage, quality-control and delivery. As stated in a 2010 UN Pamphlet, *'Human rights [to water and sanitation] do not require a particular model of service provision. They do not exclude private provision (including privatization).'* (De Albuquerque, 2010).

1.7 GROUNDWATER QUALITY – HEALTH AND CONTAMINATION REGULATIONS

The US has done much since the 1972 Clean Water Act to improve the quality of water resources and to reduce point-source contamination. However, the US Environmental Protection Agency and many environmental watch-dog NGOs claim that 40% of US rivers can be still be considered as polluted. In addition to industrial and agricultural chemicals impacting aquifers, xenobiotics and low level endocrine-disrupting compounds are an emerging threat to both groundwater and surface-water systems. A US Geological Survey study of groundwater quality showed that more than 20% of untreated water samples from 932 public wells contained at least one substance at levels of potential health concern (Toccalino *et al.*, 2010). Naturally occurring contaminants, such as radon and arsenic, accounted for about three-quarters of contaminant concentrations greater than human-health recommended limits in untreated source water. Manmade compounds found in untreated water from the public wells included herbicides, insecticides, solvents, disinfection by-products, nitrate and gasoline chemicals. These chemicals and compounds, detected in 64% of the samples, accounted for about 25% of contaminant concentrations greater than human-health thresholds.

Drinking-water related health regulations have an impact on defining which aquifers are useable for drinking water. The effect of stricter regulations can be to add more demand stress to aquifers that have no quality issues. Changing the 'rules' can have a big impact on treatment costs and on source-water management decisions. For example, changing the arsenic threshold from 50 ppb to 10 ppb (approximately 50 µg/ℓ to 10 µg/ℓ) in 2006 initiated additional annual utility costs estimated at between US$181 million and US$600 million (Tiemann, 2007). As analytical techniques become more sophisticated and health studies identify additional potential threats, the regulatory thresholds will become stricter. Federal limits for perchlorate, a man-made chemical that is used to produce rocket fuel, fireworks, flares and explosives are expected by 2013, and a decision about the regulatory levels for hexavalent chromium (chromium VI) and for fluoride are currently under review. While impacting source options, more stringent limits have provided a great stimulus to the water-treatment industry. A report from the Union Bank of Switzerland (UBS), cited in Maxwell (2011) shows an annual increase of 15% to 20% in revenues for membranes, and membrane bioreactor companies, and an average increase of 10% in revenues for companies manufacturing equipment for ozonation, reverse osmosis and ultraviolet disinfection.

In addition to federal regulations, each US state and municipal jurisdiction has regulatory authority. Each state may have its own concerns with a regulatory response that may be different or absent in other places in the state or in adjoining states. For example, the City of Austin, Texas banned the use of coal-tar-based asphalt sealants in 2006 because of pollution risk from runoff containing Polycyclic Aromatic Hydrocarbons (PAHs). Washington State also banned these types of sealants in May 2011 and a growing number of cities, now aware of the issue, are preparing to implement restrictions. Part of the challenge for professionals involved with water is the complexity of the jurisdictional layers. For example, the US has 40 federal agencies and entities with authority over water issues involving 6 cabinet departments, 13 congressional committees, and 23 congressional subcommittees. Attempts to streamline water policy have been developed as Integration of Agency Guidelines (CEQ, 2009) and are intended to promote the use of best science, peer review, full transparency and a more rigorous study process to inform authorisation and funding decisions for federal projects.

1.8   HYDRAULIC FRACTURING OF OIL AND GAS SHALES

An emerging groundwater issue in the US is that of environmental challenges to water supply associated with the use of hydraulic fracturing techniques in the development of oil, shale gas and coal-bed methane. The process uses considerable volumes of water to which various chemicals, compounds and proppants are added and pumped under pressure via long horizontal bores to liberate oil or natural gas from the rock formations. Thirty two of the USA's 50 states have reserves of oil, natural gas or coal and the drive to energy independence, and the effectiveness of hydraulic fracturing techniques, have made a huge difference to energy extraction economics in geological areas such as the Barnett Shales in Texas, the Marcellus Shales in Pennsylvania and New York and in areas such as the Powder River Basin in Montana and Wyoming. The problem is that hydraulic fracturing, which may be performed multiple times during the life of the well, requires water, and when the gas returns, can produce toxic wastewater impacted by rock chemistry and complex additives. The rapid development of gas wells – there are now over 500 000 of them in the US – has overwhelmed local treatment capacity in many places and has resulted in reports of groundwater and surface-water contamination.

The US needs energy sources and sustainable water-supply sources. The public is becoming increasingly concerned about the perceived imbalance between short-term and long-term benefits. The challenge is for energy development to proceed without jeopardy to water-supply integrity. The water-well industry's future is dependent on maintaining aquifer integrity, and groundwater professionals are therefore generally supportive of regulatory measures and protocols to minimise risks from the complex mixtures of chemical compounds used in the extraction process. These may impact the hydrological system via surface discharge or from wastewater disposal via sub-surface injection. One indirect benefit from the development of in-country gas reserves is that it further marginalises the benefits of bio-fuel crops which are extremely water intensive per (BTU) unit of heating produced.

Hydrogeological specialists in all countries where hydraulic fracturing processes are used or contemplated have a strong vested interest in working with petroleum and gas geologists to devise drilling, extraction and operation strategies that will not lead to groundwater contamination.

1.9   RECHARGE OF AQUIFERS

In the US there is a trend of 'green infrastructure' in suburban and urban areas, which involves capturing rain where it falls, and filtering out pollutants on site. Tools such as green roofs, permeable pavements, disconnecting roof drains, retrofitting detention basins, and creating rain gardens are able to reduce stormwater flow to rivers and in many cases serve to enhance and increase groundwater recharge. On a much larger scale, a slowly growing arena for groundwater expertise, with huge potential for expansion, involves aquifer storage projects. The past decade of research and operation of aquifer storage and recovery projects has seen advances in understanding the mobilisation of metals, the degradation of disinfection by-products and the attenuation of microorganisms. Several US aquifer storage and recovery (ASR) facilities have been in continuous operation for over 40 years. Currently there are well over 100 ASR projects in operation, principally in the mid-Atlantic, the northwest, Arizona, California, Nevada, Texas and Florida. The US Environmental Protection Agency estimates that the US has approximately 1 200 recharge and ASR wells that are operating or capable of operation (EPA, 2010). The conjunctive use of surface water and groundwater enables water managers to seasonally maximise economic and environmental benefits of a more dependable supply. Finding seasonally available, natural supplies of source water for recharge can be more of a challenge than deciding on the technical issues of storage and recovery.

Operational strategies, well design and source water-treatment technologies for recharge have become more sophisticated. Despite these advances, a major question for recharge proponents and experts is the need to overcome regulatory hurdles and resistance from traditional-thinking water managers who are less certain about subsurface storage they cannot see. While a major constraint on adoption of aquifer storage in the US is regulatory uncertainty, it is likely that the costs of surface infrastructure to provide for ever growing demand will provide increased focus on the lower costs and incremental spending benefits of developing projects that involve groundwater recharge.

Despite the fact that aquifer recharge and recovery is a successful water-management technology and is supported by an enormous amount of research and case studies, there are reluctant water managers who need clarity on recharge availability, potential hydrogeochemical changes during storage and ownership rights of recovered water before proceeding to invest private or public capital in strategic plans involving aquifer storage. Future expansion of recharge/recovery projects serving needs such as utility supply, irrigation, ecological maintenance, suppression of salt-water intrusion and water banking will most likely depend on the success of explaining the technology to the public, regulators and policy-makers. Indirect water reuse and stormwater capture for recharge are the only available sources in some areas to meet supply demand, prompting needs for predictions of hydrological and hydrochemical impacts. Indirect water reuse has been practiced for years in river systems with multiple users as water moves downstream. Direct water reuse means keeping the water 'captive' in an engineered environment without releasing it to spend time in the natural environment of the hydrological system (river, lake, groundwater). Storing reuse water for recharge in a natural aquifer before pumping to supply is indirect reuse. All aquifer-recharge projects, or potential projects, require input from groundwater specialists because of the need to identify and accurately characterise the receiving rock units in terms of storage potential, mobility of stored water, potential for chemical changes during storage and recovery and recharge rate calculations.

Figure 1.1 illustrates a decision-tree for potential aquifer-recharge projects for which hydrogeological criteria are critical for stop-or-go decisions related to project feasibility.

The Aquifer Recharge Decision Tree is an adaptation from work by Jen Woody (Woody, 2010) who created a similar figure related to recharge-permitting decisions. The recharge decision tree simplifies elements of the decision-making sequence and illustrates the importance of hydrogeological expertise in providing quantified assessments to the process. Recent examples from California, Florida and Washington illustrate a few of the applications of aquifer-recharge technology.

Los Angeles, the second largest city in the US, is moving forward with wastewater recycling in order to become more independent of imported water from Northern California and the Colorado River. A new US$700 million system to recycle sewage into drinking water was announced in June 2011. The groundwater replenishment

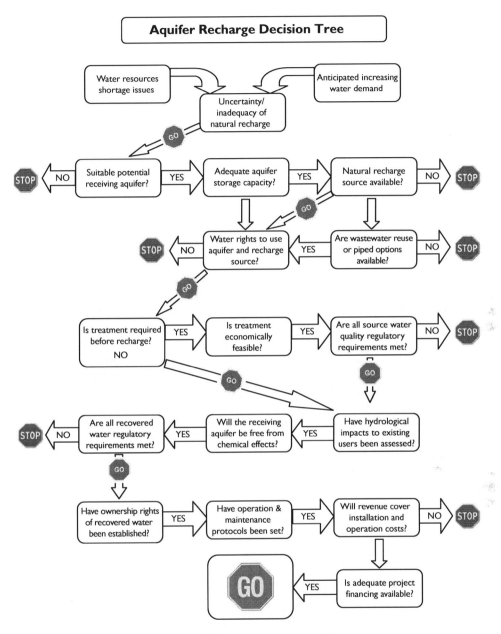

Figure 1.1 Aquifer Recharge Decision Tree.

system will be an indirect 'potable reuse' project initially purifying up to 37×10^6 m³ (30 000 acre-feet) of treated water a year which will be recharged to aquifers.

In Florida, 23×10^6 m³ (6 billion gallons) per day of fresh surface water enters the ocean (lost to tide) from south Florida (Mirecki, 2011). Population growth, salt-water intrusion, and ecosystem restoration requirements are expected to result in increasing demands on the freshwater supply. Therefore, increasing storage capacity to capture this 'lost resource' is critical for sustainable, multipurpose water supply. Above-ground reservoirs have cost and environmental disadvantages. If constructed on former agricultural lands the site footprint will require soil remediation. In addition, there are high costs of real estate, construction, operation and maintenance, and resource loss to evaporation.

A variant of recovering recharged water as a supply source is Thermal Storage and Recovery (TSR), whereby cool-temperature winter river flows are recharged and then recovered in summer to take advantage of their cool thermal properties. By using TSR, a paper-manufacturing plant on the Columbia River in the Pacific Northwest will save half a million dollars a year in energy that was previously used to cool water used in the summer months in paper manufacture (Lindsey *et al.*, 2011).

1.10 GROUND SOURCE HEATING AND COOLING (GEOTHERMAL)

The growing trend of ground source heating and cooling installations in North America (incorrectly but ubiquitously called geothermal) has provided a huge boost to the well-construction industry and is a developing field for hydrogeologists in working with engineers to characterise subsurface heat-transfer capabilities. Geothermal is a very efficient heating and air-conditioning system that has now been installed in over a million homes, schools, churches, businesses and institutions in the US. The installation costs have an investment pay-back by reducing heating and cooling costs between 50% and 80%. Many water-well drillers now have geothermal installation as the mainstay of their business.

The technology generally involves drilling vertical bores into which closed loops are inserted and then the drilled holes are back-filled with thermally enhanced grouts which maximise the transfer of heat from (or to) the adjacent geology and the fluid circulating in the loop to the heat-pump unit in the building. Some of these installations for large projects such as colleges and schools can involve 200 bores to 300 bores drilled up to 200 m deep over an area of several hectares. While mechanical engineers typically select the overall design of these installations, they are very much dependent on accurately characterised aquifer and geological information, particularly if the heat dissipation in the subsurface is dependent on groundwater flow. A concern is the potential build-up of heat in the subsurface which will progressively decrease heat transfer efficiency. Designs anticipate this potential problem by wider bore spacing and/or greater bore depth.

Already common in Europe is the technology for Aquifer Thermal Energy Storage (ATES). These systems operate by running groundwater through a cooling tower in winter and putting it back into the aquifer for storage. In summer, the chilled water is withdrawn, used for air conditioning and put back into the aquifer as warm water for use in winter to reduce heating costs. There are essentially two separate thermally modified aquifers, one used to enhance heating in winter and the other to enhance cooling in summer. Stockton College in New Jersey has this system installed. When the technology eventually 'takes-off' in the US there will be a great need for hydrogeological expertise to provide information for ATES designs.

1.11   CONCLUSION – EDUCATION AND COMMUNICATION

A recent innovation in the education process over water-policy issues is the use of computer-based informatics. This technique of information processing is a systematic approach to reviewing water-policy decisions, to identify what has been done, and what did or did not work in previous policy actions to ensure continuity and to improve policy-making and planning.

Public meetings of stakeholders concerned about a groundwater issue can be difficult to manage and the results of discussion difficult to quantify. The use of handheld wireless devices, also known as keypad polling, can transform such informational meetings by allowing participants to 'vote' anonymously and transmit choices to a laptop computer that tabulates the responses and produces graphs of the aggregated data. Such active involvement brings focus to discussions and will assist policy-makers with a clear statement of public opinion. The value of the polling process is predicated on the participants having good information on which to base their opinions. Once again, groundwater professionals have an opportunity and a challenge to provide scientific information at public meetings in a simplified format.

As water-resource decisions become increasingly bound together with economic, energy, ecological and planning decisions, groundwater professionals need to ensure that any policy involving or impacting water resources is based on good science. Groundwater specialists should take the lead in technology transfer endeavours among scientists, end-users, regulators and elected officials. Political decision-makers need help to navigate the large amounts and diverse sources of information and misinformation. With citizen organisations and environmental NGOs active as watch-dogs, it is important for them to know the basics of hydrological cause-and-effect, and more importantly, to understand where to find objective information sources. Professional associations and scientific organisations that represent groundwater interests can provide a valuable service to their members by providing educational ammunition in the form of technical explanations designed for public audiences. In fact, failure to do so may set back the implementation of groundwater projects because of perpetuated misunderstanding about the central role that natural and enhanced groundwater resources can play in water management.

REFERENCES

Bates, B.C., Kundzewicz, Z.W., Wu, S. & Palutikof, J.P. (eds.) (2008) *Climate change and water.* Technical Paper of the Intergovernmental Panel on Climate Change. Geneva, Switzerland, IPCC Secretariat.

CEQ, Council on Environmental Quality (2009) *Principles and guidelines for water and land related resources implementation studies.* Washington DC, The White House.

De Albuquerque, C. (2010) *Pamphlet, the human right to water and sanitation.* Geneva, Switzerland, United Nations Office of the High Commissioner for Human Rights.

EPA (2010) *Aquifer Recharge (AR) and Aquifer Storage & Recovery (ASR).* US Environmental Protection Agency. [Online] Available from: http://water.epa.gov/type/groundwater/uic/aquiferrecharge.cfm

Fishman, C. (2011) *The Big Thirst – The Secret Life and Turbulent Future of Water.* New York, NY, Free Press.

Johnson Foundation (2010) *Charting new waters: A call to action to address US freshwater challenges*. Racine, WI, Johnson Foundation at Wingspread.

Lindsey, K., Barry, J., Augustine, C., Lam, R. & Tobin, D. (2011) Aquifer recharge & thermal storage recovery. In: *American Ground Water Trust Aquifer Recharge Conference*. Conference Presentation, Olympia, WA.

Maddaus, W.O. (2008) *Integrated resource planning in the new AWWA M50 Manual*. Denver, CO, American Water Works Association, Sustainable Water Resources.

Maxwell, S. (2011) *Water market review*. Boulder County, CO, Techknowledgey Strategic Group. p. 24.

Mirecki, J. (2011) Conjunctive use of aquifer storage recovery and reservoirs for cost-effective water resource management in the comprehensive everglades restoration plan. In: *NGWA Ground Water Summit*. Conference Presentation, Baltimore, MD.

Tiemann, M. (2007) *Arsenic in drinking water*. Washington DC, Congressional Research Service, Library of Congress. Report to Congress – Order Code RS20672.

Toccalino, P.L., Norman, J.E. & Hitt, K.J. (2010) *Quality of source water from public-supply wells in the United States, 1993–2007*. U.S. Geological Survey Scientific Investigations. Report 2010–5024, p. 206.

Woody, J. (2007) A preliminary assessment of hydrogeologic suitability for aquifer storage and recovery (ASR) in Oregon. *Master's Thesis*, Corvallis, OR, Oregon State University, p. 325.

Chapter 2

The challenges of municipal groundwater supply from private land

H. Seyler, A. Mlisa & K. Riemann
Umvoto Africa (Pty) Ltd, Muizenberg, Cape Town, Western Cape,
South Africa

ABSTRACT

According to the National Water Act a landowner does not own the water under his land. The use of groundwater from private land for municipal domestic supply is therefore possible and promoted. The National Water Act allows the Department of Water Affairs the right to access private land for monitoring purposes, but this right is not extended to municipalities as Water Services Authorities. The National Water Act gives the municipality the right to claim a servitude over private land in order to realise a water use authorisation (i.e., after being licensed), but the municipalities must apply the Water Services Act in order to lay services on private properties. Some municipal ordinances (laws passed by a municipality or local authority) state that the only compensation required for the laying of any service, is that which is calculated based on the value of land under the servitude area. However, no legislation overtly allows a municipality to gain entry for groundwater exploration or drilling, prior to laying of the service. The problems encountered and the hurdles overcome in groundwater development from private land for municipal supply are discussed.

2.1 INTRODUCTION

Groundwater has historically been given limited attention, and has not been perceived as an important water resource in South Africa. This is reflected in general statistics showing that only 13% of the nation's total water supply originates from groundwater (DWA, 2008a). Public perception remains that groundwater is not a sustainable resource for bulk domestic supply and cannot be managed properly. Despite this, a growing number of municipalities utilise groundwater on a regular basis, and provide examples of successful management of this resource.

A number of guidelines for groundwater management have been developed internationally and for the South African context, these include the NORAD toolkit (DWAF, 2004a), the Water Research Commission (WRC) 'Guidelines for the monitoring and management of ground water resources in rural water supply schemes' (Meyer, 2002) and the Department of Water Affairs (DWA) 'Guideline for the Assessment, Planning and Management of Groundwater Resources in South Africa' (DWA, 2008b). Other documents include water-quality management protocols, minimum standards, the Framework for a National Groundwater Strategy (DWAF, 2007a), the National Water Resources Strategy (DWAF, 2004b), the

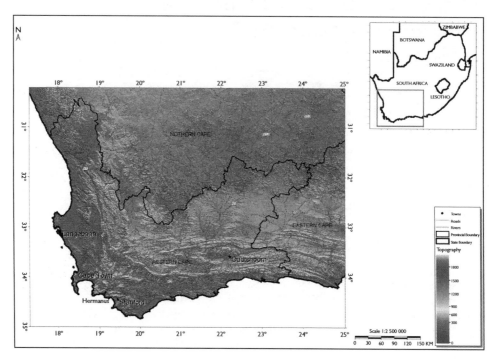

Figure 2.1 Location map.

'Guidelines for Catchment Management Strategies towards Equity, Sustainability and Efficiency' (DWAF, 2007b) and regional groundwater plans, as well as selected national and international articles and publications on groundwater management aspects.

However, none of these guidelines provide the local authorities with tools on how to deal with access to private land. This hinders groundwater development in the country and as a result, several Water Services Authorities (WSAs) have opted to only develop groundwater on municipal property, or this is seen to be the first option. This means that even when desktop studies have shown that the best groundwater yield will be on private land, the WSAs opt for exploration on municipal land first and only when it is proven that the yield will not be sufficient they commence with the process to access private land for drilling and eventually abstraction, causing project delays and increased costs. During the recent drought, municipalities in the Eastern and Western Cape, for example, (see Figure 2.1 for location map), focused the drilling of emergency boreholes on municipal land in order to expedite the supply, rather than targeting the most favourable hydrogeological sites, resulting in lower than possible yields (Parsons, 2011).

The preference for municipal land is understandable as recent groundwater-development projects in the Eastern and Western Cape have shown that the process of supply from private land, i.e. achieving landowner agreements and registering servitudes, is problematic.

2.2 LEGAL FRAMEWORK FOR ACCESS TO WATER AND WATER DELIVERY

A number of laws, from the South African Constitution to municipal by-laws, govern the access to water in South Africa. The South African Constitution supports access to water for all and gives rights to the relevant authorities to ensure delivery of water. Section 27 states that '… Everyone has the right to have access to sufficient food and water … the state must take reasonable legislative and other measures, within its available resources, to achieve the progressive realisation of … these rights'.

The South African Constitution allocates the management of water resources to National Government and the management of water and sanitation services for all citizens to municipalities (local government). Hence there is one Act that deals with the sources of water (National Water Act, national responsibility) and one Act that deals with water services (Water Services Act, local responsibility). The public water and sanitation sector in South Africa is organised in three different tiers:

- the national government, represented by the DWA, establishes policy;
- Water Boards, which primarily provide bulk water, in addition to playing a role in water resources management. They also provide some retail services and operate some wastewater treatment plants;
- municipalities, which provide most retail services and also own some of the bulk water infrastructure. The municipalities do this through WSA and Water Service Provider (WSP) departments.

The National Water Act (Act 36 of 1998) provides a framework to protect water resources against over-exploitation and to ensure that there is water for social and economic development and water for the future. It also recognises that water belongs to the whole nation for the benefit of all people. The Act outlines the permissible use of water. It states that a person can:

- take water for reasonable domestic use in their household, directly from any water resource to which that person has lawful access;
- take water for use on land owned or occupied by that person, for reasonable domestic use; small gardening (not for commercial purposes); and the watering of animals (excluding feedlots) which graze on that land (within the grazing capacity of that land) from any water resource which is situated on or forms a boundary of that land, if the use is not excessive in relation to the capacity of the water resource and the needs of other users;
- store and use runoff water from a roof;
- in emergency situations, take water from any water resource for human consumption or fire fighting.

The National Water Act (NWA) also states that a landowner does not own the water under that land (Section 24, Act 36 of 1998). Municipal abstraction from private land is therefore possible and promoted. However, even though this is not an explicit requirement in the NWA, the DWA requests that landowner agreements are presented before considering any licence application or funding for groundwater development projects on private land.

South Africa's Water Services Act (Act 108 of 1997) contains a section on the right of access to basic water and sanitation. It states that:

- everyone has a right of access to basic water supply and basic sanitation;
- every water services institution must take reasonable measures to realise these rights;
- every water services authority must, in its water services development plan, provide for measures to realise these rights.

Under the Water Services Act the municipalities are allowed to lay services on properties but cannot necessarily gain entry for groundwater exploration or drilling prior to laying of the service. According to the NWA the DWA has the right to access private land for drilling exploration boreholes for monitoring purposes (Section 125(2), Act 36 of 1998), but this right is not extended to municipalities as Water Services Authorities.

The municipal by-laws and ordinances give municipalities a certain amount of legal leeway for accessing private land, for example for the laying and maintenance of services; however, enforcement of these laws and their legal value is not clear. In terms of the municipal ordinance, the only compensation required for the laying of a service is calculated based on the value of land under the servitude area.

When drilling of exploration boreholes and abstraction of groundwater was considered a listed activity under the National Environmental Management Act (NEMA) (Act 107 of 1998), requiring a Record of Decision and an Environmental Impact Assessment (EIA), landowner consent was a requirement if the activity were to occur on the applicant's land, e.g. on private land, for a municipality as applicant. This caused confusion among landowners who considered that this would impact on their right to give comment as an interested and affected party under the EIA process, and object to the activity should they so choose, which in turn caused delays with the EIA process as basic assessment reports could not be finalised and submitted or site evaluation could not be carried out until the consent forms had been signed. The recent change of the NEMA regulation for EIA requirements with respect to groundwater development is welcomed, as it is a step towards alignment of the different processes.

2.3 GROUNDWATER DEVELOPMENT AND LANDOWNER AGREEMENT PROCESS

2.3.1 Groundwater development process

The development of groundwater schemes should follow a standard project process, which is similar to other infrastructure developments (Riemann *et al.*, 2011). The main elements of this process are shown in the first three columns of Table 2.1, and go from conceptualisation, to reconnaissance and pre-feasibility, right through to design and implementation, and operations and maintenance.

Table 2.1 Phases of groundwater development and landowner agreements (after Riemann et al., 2011).

Level of study	Product/decision	Data collection	Landowner agreement requirement, and guiding legislation
Conceptualisation Reconnaissance	Inception/Planning report ID target areas Recommendation for and prioritisation of monitoring	• expert evaluation of existing data • primarily desktop work with limited fieldwork and data collection, as required (e.g., Hydrocensus) • 1st order water balance model	• not required • access to private land for field work • no legislation in support of this
Pre-feasibility	Environmental monitoring and assessment Identify target sites	• geological and ecological mapping • installation of monitoring infrastructure and ongoing monitoring of relevant processes • re-calibrate water balance model	• access required for establishing monitoring infrastructure and ongoing monitoring • no legislation in support of this

Feasibility	Exploration
	Yield estimation
	Water Use Licence application
	EIA application (if applicable)
	Feasibility report
		• site survey, borehole siting
		• drilling and testing of exploration boreholes
		• regional groundwater modelling
		• invest in collecting all relevant input for design purposes
			• access required for borehole siting, exploration drilling and servitude or other protective agreement required for continuous monitoring
			• no legislation directly in support of this:
			  o exploration and monitoring boreholes not seen as services by the Water Services Act.
			  o DWA is supported by the NWA to establish exploration/monitoring boreholes but not the WSA for licensing purposes DWA requires the WSA to have landowner agreements in place

Options analysis	Options analysis report	• comparison of different options for water supply, based on feasibility studies	• no access required

Design and implementation	Well-field design and implementation
	Operating rules
		• design all components of the scheme
		• well-field model
			• access agreement and servitude or other protective agreement required for drilling of production boreholes and laying of services, e.g. pipeline
			  o the Water Services Act grants the Water Service Provider the legal right to lay services
			  o the Constitution and NWA provide guidance on compensation
			  o the NWA grants the right to register a servitude to a licensee

Operation and maintenance	Operation and maintenance	• ongoing monitoring	• servitude agreement required for permanent access to undertake monitoring and maintenance

As groundwater development is a phased approach, access agreements, servitudes and compensation agreements are required at different times and levels of the groundwater development project, as shown in the fourth column of Table 2.1:

- access for exploration field work from reconnaissance level onwards;
- access for installation of monitoring network from pre-feasibility level onwards;
- access for drilling of exploration and or production boreholes at feasibility and design level;
- servitude for monitoring from pre-feasibility level onwards;
- servitude for services, e.g. borehole, pipeline, pump house from design level onwards;
- compensation for servitude area and use of water from operational level onwards.

2.3.2 Access agreement

Access agreements can have many forms and details. For once-off access at an early stage of a project, a hand-shake agreement or verbal permission is normally sufficient, while the installation of equipment or drilling should only commence, once a written agreement is in place. There is no guideline or template for these agreements, and each one must be negotiated on its own merit.

The EIA and licensing processes require the consent of the owners of the land on which land the activity will occur. The Department of Environmental Affairs uses a consent form for the EIA process that simply states that the landowner gives consent to the activity being investigated and considered. This does not imply that the landowner agrees with the activity taking place and does not foreclose any comments or concerns by the landowner.

A special form of agreement is the servitude that is registered in the title deeds. A registered servitude gives the municipality the right to access the land under servitude for reasons stipulated in the servitude. The landowner still has free access to the land, but might be restricted in his activities, so that the servitude can be effective. Since the servitude is registered in the title deeds, it remains valid when selling the property.

2.3.3 Compensation

The question of compensation is addressed in the South African Constitution, Section 25, 2) '... *Property may be expropriated* only in terms of law of general application ... for a public purpose or in the public interest; and *subject to compensation*, the amount of which and the time and manner of payment of which have either been agreed to by those affected or decided or approved by a court.' It further states that '... The amount of the compensation and the time and manner of payment must be just and equitable, reflecting an equitable balance between the public interest and the interests of those affected, having regard to all relevant circumstances, including

a   the current use of the property;
b   the history of the acquisition and use of the property;
c   the market value of the property;

d   the extent of direct state investment and subsidy in the acquisition and beneficial capital improvement of the property; and

e   the purpose of the expropriation.'

The Constitution further states that for the purposes of Section 25 '... the public interest includes the nation's commitment ... to bring about equitable access to all South Africa's natural resources ....' The NWA also stipulates conditions for servitudes and associated compensations, effectively stating that compensation should be based on the value of the affected land, and therefore can only be applicable from the design and implementation stage of a groundwater development project onwards, since only at this point is there a footprint of the affected land.

However, there might be a case for compensation for inconvenience during earlier phases of the project, meaning compensation for day-to-day disturbance of the landowner by the project activities such as drilling, which mainly occur during feasibility and design levels of a project. In most cases the landowners request far more compensation than what the WSA is legally required to grant them.

Following the recommendation of the Constitution and the NWA on compensation, the minimum requirement is deemed to be the value of land to be used (area) which is a straightforward land evaluation by a municipal land evaluator. The current or potential land use and hence loss of 'economical productivity' of that land to the landowner is often not considered.

2.3.4   Discussion and examples

Only the landowner agreements required for the design level are supported by the current laws, although even those are not sufficient as they do not give the WSAs the full rights required for successful and effective implementation. To summarise the conundrum, in order to access land for exploration drilling the WSA needs a servitude, and in terms of the NWA, only an authorised (i.e. licensed) party is entitled to claim servitude. To become authorised, according to DWA, a WSA must include a landowner agreement in a water-use licence application. Hence the process comes full circle and requires landowner consent. Furthermore, registration of servitude prior to exploration is a fruitless expense should exploration show that the groundwater resource is not viable. Also, the servitude need only surround the borehole infrastructure/pipeline route, especially as compensation is based on area of land under servitude, and this pipeline route can often deviate from the initial design during construction, thus it is appropriate to register servitude based on built infrastructure.

In a situation where the groundwater resource is unknown and pre-exploration is required to determine what the exploration is worth, commencement on a handshake agreement is cost-effective. Drilling of exploration boreholes commenced in the Oudtshoorn Municipality without having landowner agreements and servitudes in place. Subsequently, one of the landowners denied access to the borehole for monitoring and used the artesian borehole for irrigation. Similarly an artesian monitoring borehole, drilled during a WRC project for the City of Cape Town on communal land, managed by Cape Nature, has been opened to be used for an ablution facility on the nearby picnic spot.

The Overstrand Municipality underwent lengthy negotiations with one landowner close to the small town of Stanford. The aim was to negotiate an access agreement for exploration – an initial geophysics survey, and subsequent to these results, the drilling and testing of exploration boreholes to obtain a rough estimation of the production potential of the borehole. Should the exploration boreholes be successful, the municipality would also want to be able to utilise these assets as production boreholes. The landowner understandably wished to protect themselves and have all final details agreed upon upfront, such as the actual value of the servitude compensation, should exploration be successful and a servitude needed to be registered against the title deeds of the property. This detail cannot be known before exploration. As the negotiations became lengthy, their worth became questionable given that the resource was not yet proven. For this reason and due to additional requests for private assets such as property fencing, and private boreholes, negotiations were eventually abandoned.

The exploration moved to a neighbouring property, at which the landowner was happy to proceed on a written Memorandum of Agreement that essentially listed the future processes that would occur if exploration was successful, and the legislation that would support each stage (i.e. the process shown in Table 2.1). Therefore the exact final servitude compensation amount was not listed and agreed upfront, yet the process for calculation of it was agreed. The exploration area was, however, limited to what the landowner considered 'dead space', that is private land between the fence line and the road reserve, and the municipality was forced to provide additional compensation which it was not legally required to pay. An agreement was reached that should the exploration be successful and the boreholes be used for production, a fixed percentage of the abstracted water was to be supplied to the landowner. This supply of water is not necessary, but was agreed upon by the municipality as the associated pumping costs were deemed to be less than the alternative, namely expropriation. Fixing the agreed volume to a percentage of the abstracted water meant that the municipality would not be penalised should the exploitable volumes be small.

Unfortunately, the entire process of landowner agreements and compensation for access of land for groundwater development is not regulated. As there is no guideline available the municipality must become a negotiator involving bargaining, handshake agreements, pleas and memoranda of agreements.

2.4   POSSIBLE WAY FORWARD

Though the Constitution and the NWA allow for expropriation, this is found to be an extreme path by all parties involved and hence it has been found that servitudes registered against the title deeds of the servient properties are the most effective option for a final agreement for access to land. The following process, which was eventually applied by the Overstrand Municipality as described above, is considered appropriate for groundwater development projects:

- signing of an 'in principle' Memorandum of Agreement prior to commencement of exploration drilling, i.e. at feasibility level – reaching an 'in principle' agreement with the landowners over whose property access for drilling is required.
  - ◦ if possible, the Memorandum of Agreement should cover all potential future activities required to be undertaken on the land, including the design level;

- ◦ the Memorandum of Agreement should indicate that a subsequent servitude registration might occur depending upon the results of the feasibility and design levels of the study;
- ◦ agreement should consider the duration of the intended servitude, i.e. will it be for a fixed period of time or will it subsist indefinitely or be based on activity;
- ◦ the Memorandum of Agreement should indicate that compensation for the servitude will be negotiated once the results of the design level are available, meaning once it is known whether the boreholes will be used for abstraction of water. It is possible to include two scenarios for the compensation, one in which the boreholes are used only for monitoring and one in which they are used for future production.

- • servitude to be registered by way of an endorsement/restriction on the title deeds for the property concerned so that it gives rise to what is known as a real right, i.e. it is enforceable against the world at large. This would protect the municipality in a situation where the owner of the land concerned sold the property, as the new owner would be bound by the servitude as reflected on the title deeds of the property.

Following the above two-step process ensures that the negotiations for compensation are carried out once the WSA has established the value of the water found underground on the land and hence can evaluate a reasonable compensation for the access and use of water. The request by many landowners for use of water as compensation is premature before production boreholes have been drilled and tested as the amount of compensation cannot be agreed upon until the yield of the boreholes has been established. Agreeing to a percentage or proportion of water abstracted undoubtedly protects the WSA. Commencing exploration on the basis of verbal consent is cost-effective in terms of negotiation time but the lack of written consent enables the landowners to refuse access to the established boreholes, which inhibits the monitoring process; for example, two years' worth of monitoring data were lost in a project, whilst access was being renegotiated. Hence, the 'in principle' Memorandum of Agreement is a strongly suggested step prior to drilling.

The negotiation and agreement with respect to the compensation is a lengthy process, if proper guidelines are not followed. The minimum requirement of the land value is often not accepted by the landowners, while their requests for special items and assets to be included, for example fencing, an additional private borehole, free water supply, cannot be entertained by the municipality. However, it has been shown to be a very successful approach to 'buy' the water from the landowner for a value similar to the potential profit, if the landowners were to use the water on their property for economic benefit.

In estimating reasonable compensation, the WSAs have to consider the following additional costs related to servitude registration:

- • surveying of the section of the property to be registered for servitude, including dimensions of the road access;
- • survey diagrams lodged with and approved by the surveyor-general and, thereafter, the servitudes would have to be formally registered against the properties' title deeds at the deeds office;

- land valuation fees; it is preferable to use a municipal valuer although he/she should be appointed by mutual agreement between the parties to avoid perception of biases;
- lawyers' costs for all parties involved, including the lawyers for the landowners. In this case the WSAs are advised to set a maximum charge per hour and expected time to be spent by the lawyer.

2.5   RECOMMENDATIONS AND CONCLUSIONS

Although the Constitution, NWA, the WSA and the municipal by-laws and ordinances support the municipal procedures and allow for servitude for water works with agreed compensation, a key challenge for the municipalities, the landowners and the DWA is to run the various legal processes in parallel whilst following the correct process in dealing with requests and concerns.

Although the NWA Section 24 requires the consent of landowners for the use of water found underground on that land, it is suggested that the DWA should consider water-use licence applications without having the requirement for landowner agreements in place. A landowner's concerns regarding private water supply can be dealt with in the comments and responses register of the public participation during the licensing process, and the water could be allocated to the municipality without the consent. The passing of a licence then allows the municipality to establish the water works and lay the pipeline. It is further suggested that the right for the establishment of exploration boreholes and the access to land for monitoring purposes be extended to include the WSAs and WSPs.

ACKNOWLEDGEMENTS

This paper draws on lessons from two projects by Umvoto Africa Pty (Ltd) in which the support of the Overstrand municipality and the Oudtshoorn municipality is acknowledged.

REFERENCES

Department of Water Affairs and Forestry (2004a) *A Framework for Groundwater Management of Community Water Supply.* The NORAD-Assisted Programme for the Sustainable Development of Groundwater Sources under the Community Water and Sanitation Programme in South Africa. Pretoria, South Africa, DWAF.

Department of Water Affairs and Forestry (2004b) *National Water Resource Strategy.* 1st edition. Pretoria, South Africa, DWAF.

Department of Water Affairs and Forestry (2007a) *A Framework for a National Groundwater Strategy (NGS).* 1st edition. Pretoria, South Africa, DWAF.

Department of Water Affairs and Forestry (2007b) *Guidelines for Catchment Management Strategies towards Equity, Sustainability and Efficiency.* 1st edition. Pretoria, South Africa, DWAF.

Department of Water Affairs (DWA), South Africa (2008a) *Strategic Framework on Water for Sustainable Growth and Development – Summary Discussion Document*. Pretoria, South Africa, DWA.

Department of Water Affairs (DWA), South Africa (2008b) *A Guideline for the Assessment, Planning and Management of Groundwater Resources in South Africa*. 1st edition. Pretoria, South Africa, DWA.

Government of South Africa (1997) South African Water Services Act. Act No. 108 of 1997. *Government Gazette* Volume 390.

Government of South Africa (1998) South African National Water Act. Act No. 36 of 1998. *Government Gazette* Volume 398.

Government of South Africa (1998) The National Environmental Management Act. Act No. 107 of 1998.

Meyer, R. (2002) *Guidelines for the monitoring and management of ground water resources in rural water supply schemes*. Pretoria, South Africa, Water Research Commission. WRC Report No. 861/1/02a.

Parsons, R. (2011) Experiences from developing emergency groundwater supplies for the drought-impacted Southern Cape Region. In: *Proceedings of the Biennial Conference of the Groundwater Division of the Geological Society of South Africa, CSIR International Convention Centre, Pretoria, South Africa. 19–21 September 2011.*

Riemann, K.D., Louw, N., Chimboza, N. & Fubesi, M. (2011) *Groundwater management framework*. Pretoria, South Africa, Water Research Commission. WRC Report No. 1917/1/11.

Chapter 3

When are groundwater data enough for decision-making?

J.J.P. Vivier[1] *& I.J. Van Der Walt*[2]
[1]*AGES Consulting, Pretoria, Gauteng, South Africa*
[2]*University of the North-West, Potchefstroom, North-West, South Africa*

ABSTRACT

In groundwater studies, especially complex environmental geohydrological investigations, there is a tendency for analysts and reviewers to require more data. The question is whether more data are actually better? The role of data and information in the decision-making process was evaluated, with the aim of characterising the decision process in terms of the influence of data and information in that process. When data are analysed, information is generated which upon interpretation increases the level of knowledge and understanding that is used to base management decisions on. To determine the effect of data and information on the decision-making process, data from a field site were evaluated. The depth-to-groundwater level was used as the required variable on which information was required. The evaluation was done for all 715 data points collected for the Middelburg Site aquifers. The analysis showed that the decision-making process has a logarithmic nature which means that less and less information becomes available with more data. The value of data was assessed in a data-worth or data-cost evaluation. The information gained and cost can be determined when sufficient and when optimal data have been gathered. More data are not better from a decision-making and cost perspective.

3.1 INTRODUCTION

Complex environmental groundwater-management problems have been created in South Africa, due to mining and industrial operations; this situation is exacerbated by water decanting from abandoned gold mines (Handley, 2004). These problems have escalated to the level where decisive action needs to be taken, which poses a challenge for decision-makers. Environmental water-management investigations are done based on scientific methods, which in most cases identify inadequacies in data and information that highlight the uncertainties associated with these problems. The result is that additional data is usually required to scale up the scientific investigations in an attempt to reduce uncertainties. This leads to a requirement of more investigations where even more data gaps or complexities of the problem are identified. In these cases, where an assumed data deficiency is perceived by analysts, management objectives cannot be met as present methods of assessment require increasing detailed data sets that are seldom available, rather than supporting the decision-making process. This leads to a divergent process that highlights uncertainty and counteracts the decision-making process.

In this paper, the role of data and information in the decision-making process is evaluated to characterise the process. To evaluate whether more data are better, a groundwater case study is presented as an example of an environmental component.

3.2 OBJECTIVE

The objective of this investigation was to determine the following:

- Do more data equate to better information?
- What is the ideal state of data and information in decision-making? Is the ideal state achievable?

3.3 DATA AND INFORMATION IN THE DECISION-MAKING PROCESS

The basis of any decision process whether quantitative or subjective, is *information*. Information is based on *data*[1] and the quantity and quality of the data will influence the quality of information. Data and information are therefore the building blocks of the decision-making process. It is important to consider the role or influence of data and information in the decision-making process. In this study, the adequacy of data to address groundwater problems was investigated, to determine when sufficient information is yielded for the purposes of decision-making.

When data are analysed, information is generated which upon interpretation increases the level of knowledge and understanding that is used as the basis for management decisions (Figure 3.1). The purpose of data collection, e.g., a groundwater assessment such as to determine the Reserve[2] must be to assist in decision-making such as on resource management. This process requires that the resource must be quantified. Resource quantification provides decision-making information and depends on the availability of data. For instance, if excess water is available in an aquifer following the allocation to the Reserve components, there is no requirement to protect the resource other than stating what volume of groundwater can still be used or allocated for development and vice versa.

3.4 DATA AND INFORMATION IN GROUNDWATER: A CASE STUDY

The aim of this section is to determine whether 'more data are better' based on data gathered from field sites. Groundwater case studies from three sites namely; The Middelburg Site (AGES, 2010c), Kalahari (AGES, 2010d) and Sand River (AGES, 2010e) Sites were used where boreholes were drilled to characterise aquifer systems (Figure 3.2).

---

1 In general, raw data that (1) has been verified to be accurate and timely, (2) is specific and organized for a purpose, (3) is presented within a context that gives it meaning and relevance, and which (4) leads to increase in understanding and decrease in uncertainty. The value of information lies solely in its ability to affect a behaviour, decision, or outcome. A piece of information is considered valueless if, after receiving it, things remain unchanged (www.businessdictionary.com).

2 The water Reserve is legislated in the National Water Act (NWA, Act 36 of 1998) as the basic human need and environmental water requirements that have preference and which must be allocated before other uses are considered.

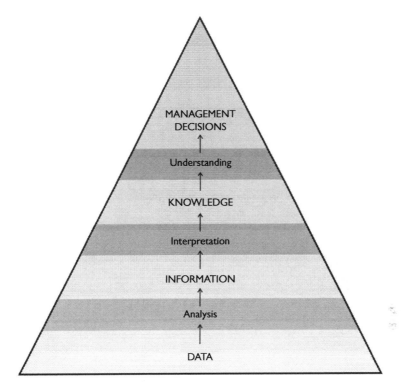

Figure 3.1 The role of data in the decision-making process (own construction).

The Middelburg Site will be described in this paper. When a new aquifer system is developed, where say no prior information is available, new boreholes are drilled to characterise the aquifer. Let's consider that the regional depth-to-groundwater level is the variable that is required to be analysed (it could be any other variable such as transmissivity, etc.). This analysis can be much more complex if, for example, impacts of abstraction on water levels are determined, etc. To keep it simple, the depth to groundwater on a regional scale was evaluated.

The first borehole is considered to provide the most information as no prior information is available. The second and third boreholes would continue to do the same and so on. It must be established whether the data that are gathered lead to *sufficient* information at some point for the purposes of decision-making (Figure 3.3). Information can be defined here as an accumulation and arrangement of data (i.e. cumulative data). Data is seen as analogue to pixels and information is the picture that is formed from the accumulation and arrangement of the data pixels. For this purpose, the field-site data were considered for existing boreholes and new boreholes were drilled to determine the depth-to-groundwater level as the unknown or uncertain parameter.

It is considered that information regarding the depth to groundwater is determined by the average and the standard deviation. If the additional data points do not contribute significantly to the change in the average or the change in the standard deviation, then

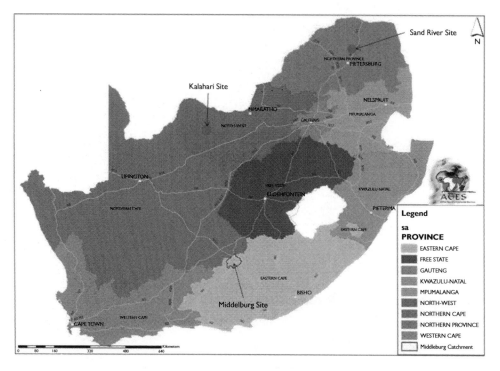

Figure 3.2 Location of the three field sites.

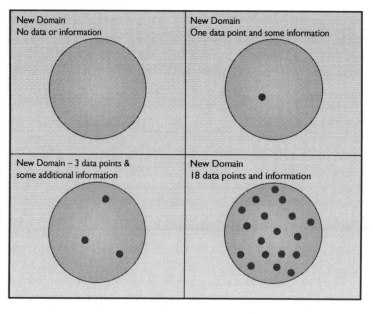

Figure 3.3 Schematic representation the data gathering process in a new, unknown domain (own construction).

the information for the area is considered as *sufficient*. *Perfect data* are then defined as data with a standard deviation of zero, which like in the case of *perfect information* does not exist in nature. If the standard deviation would be zero, the analyst would require only one data point to obtain perfect information and would have no uncertainty in the data, as all the points would have the same value. Information is represented by the change in, for example, the average depth to groundwater for each additional accumulative data point that is gathered. If this value does not change significantly, then no significant information is gained, despite the fact that more data is gathered.

To investigate the role of data in painting the information picture, the Middelburg Site was evaluated. The Middelburg study area covers three quaternary sub-catchments (Q14A, Q14B and Q14C) in the Eastern Cape, covering a surface area of 2052 km² (Figure 3.2; Figure 3.4). The purpose of the project was to supply groundwater to the town of Middelburg in the Eastern Cape Province (AGES, 2010c). The depth to groundwater is an important parameter as it indicates whether the groundwater resource is over-utilised or not.

The water-level data were obtained from a regional hydrocensus during which 715 boreholes were monitored (Figure 3.4). The first borehole surveyed at the Middelburg Site indicated a water-level depth of 12.88 m. The second borehole had a water-level depth of 12.26 m with an average between the two points of 12.57 m. The third borehole had a water-level depth of 12.42 m with a cumulative average of the three points of 12.52 m. The fourth borehole had a water-level depth of 6.47 m; this value changed the cumulative average between the first four points to 11.01 m, and so on. A graph of the 715 data points (Figure 3.4; Figure 3.5) shows that the cumulative

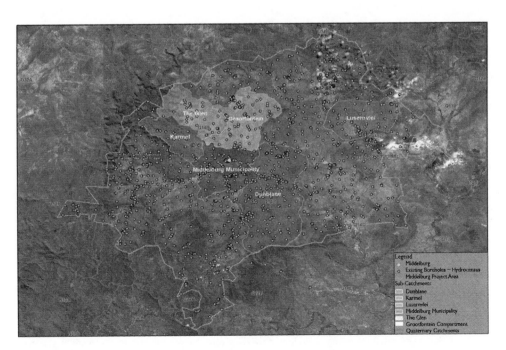

Figure 3.4 Middelburg – catchment map showing the borehole distribution (AGES 2009).

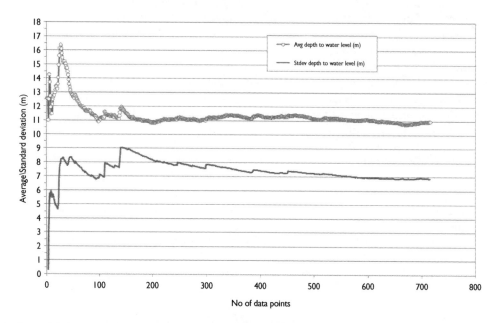

Figure 3.5 Middelburg Site depth-to-water level: Graph showing the cumulative average and cumulative standard deviation with increasing data points.

change in the average and the standard deviation converges with the number of data points (Figure 3.5). As it is accepted that perfect information is not attainable, the analyst must decide on a maximum error or convergence value that would be considered as *acceptable* or *sufficient*. In the case of the Middelburg Site, the average depth to groundwater converges towards 10.95 m and the standard deviation towards 6.9 m after 715 data points. The coefficient of variation (COV)[3] is high at 63%, which indicates a high variability and hence high uncertainty associated with the data type.

To determine the change in the variables with the number of data points, the first derivative was determined for both the change in the cumulative average and the change in the cumulative standard deviation with an increasing number of data points. The trend shows an exponential decrease in change with an increase in the number of data points (Figure 3.6). Except for some outliers, after six data points, the change in the average water-level value is effectively less than 1 m (representing an error of <9% of the actual average of 10.95 m). The change in the standard deviation is smaller than 1 m after 25 data points and smaller than 0.5 m after 30 data points (Figure 3.6). An error below 0.1 m would require more than 500 data points.

The analyst may also use the error determined as a percentage of the cumulative value, for a decision point as to when *sufficient* data have been gathered (Figure 3.7).

---

3 The COV is a measurement of the variability of the data and represents the standard deviation as a fraction of the average (http://www.businessdictionary.com/definition/coefficient-of-variation.html).

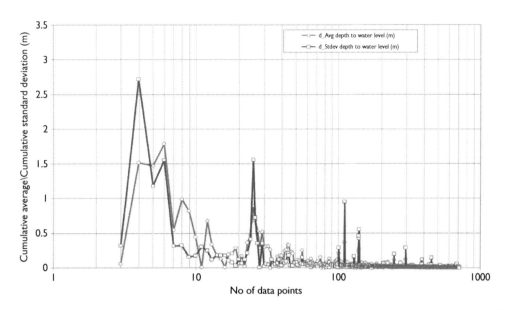

Figure 3.6 Middelburg Site depth-to-water level: Semi-log graph showing the change in the cumulative average and the change in the cumulative standard deviation with increasing data points.

Consideration of the percentage error relative to the average after 715 data points would indicate when the error is, for example, smaller than 10% or 5%. An error of less than 5% is reached after 25 data points have been gathered. Although it approaches zero, it will never reach it (Figure 3.7).

The trend shows that exponentially less information is provided with an increase in the number of data points (Figure 3.8). The information trend is similar to the law of diminishing returns used in economics (Mohr and Fourie, 2004). It forms a very good correlation ($R^2 = +0.95$) with a logarithmic trend on a linear plot and straight line on a semi-log plot (Figure 3.9) that is defined by:

$$y = a \ln(x) - b \qquad (3.1)$$

where:

    a is the number of data points

    b represents the information index.

The *information index* is defined by an *accumulation of data* on the y-axis and converges towards a maximum value (Figure 3.8; Equation (3.2)). This value represents the cumulative change in the average and the cumulative change in the standard deviation of the depth-to-groundwater level and hence the information that is provided by consecutive data points (Figure 3.8; Figure 3.9). Data points that do not significantly change the average value or the standard deviation of the variables do not increase the level of information and knowledge about the variable in consideration.

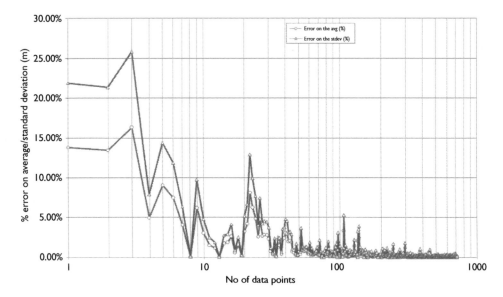

Figure 3.7 Middelburg Site depth-to-water level: Semi-log graph showing the percentage error relative to the actual value.

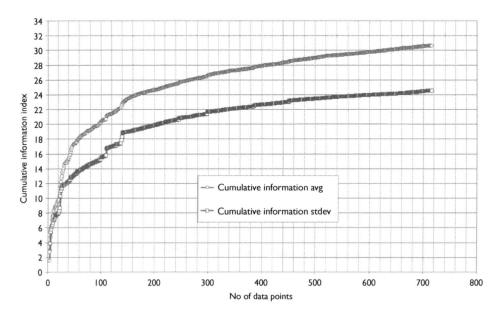

Figure 3.8 Middelburg Site depth-to-water level: Cumulative information graph.

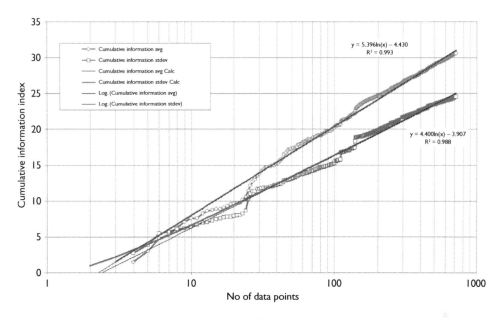

Figure 3.9 Middelburg Site depth-to-water level: Semi-log plot of cumulative information graph.

Perfect information[4] can now be defined as data with an error of zero and hence zero uncertainty, which like in the case of *perfect data* does not exist. It does, however, provide a limit that can be used as a reference.[5] Information was earlier defined as an accumulation of data. It can be determined mathematically as the cumulative change in the variable represented by the first derivative:

$$I_{ind} = \sum_{i=1}^{n} \frac{dv_i}{dn_i} \qquad (3.2)$$

where:

n are the number of variables v from i to n

I_{ind} represents the information index (Figure 3.8)

The information index therefore represents the difference in the average water level determined from the first 2 data points, then the first 3 data points and so on (Equation (3.2)).

Perfect information is reached when the gradient of the information curve becomes zero (i.e. a horizontal line), which will not happen in practice. The semi-log plot of the information curve shows that it assumes a straight line (Figure 3.9). The identification

---

4 Perfect information is defined as an illustrative reference of when information is 100% accurate. It would be represented by a horizontal line on the linear trend plot (Figures 3.8).

5 The same as the principle of an ideal gas is used in physics to characterize gases (Beuche 1986).

of a straight line on a semi-log plot of the first derivative could serve to identify when data become *sufficient* information.

The cumulative line can be called the *information curve*, which converges to a value of 35 on a linear plot (Figure 3.10). As discussed earlier, the shape of the information curve indicates that less and less is gained in terms of information as the number of data points increases (Figure 3.8; Figure 3.9). Apart from the fact that the analyst may choose a minimum error in terms of the actual value (Figure 3.6) or as a percentage of the expected value (Figure 3.7), the question arises, when are data *sufficient, near-optimal* or *optimal*? To evaluate this, four straight gradient lines were fitted to the semi-log plot to determine the change in information with the increase in the number of data points (Figure 3.10). The aim is to determine when additional information becomes insignificant if compared to the number of data points already considered. In other words, the analyst wants to determine when no more significant information or knowledge is gained about the variable under consideration.

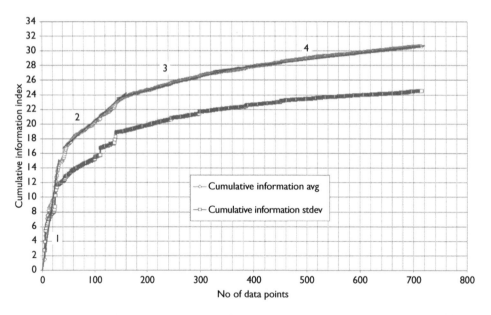

Figure 3.10 Middelburg Site depth-to-water level: Cumulative information graph with straight lines fitted.

Table 3.1 Middelburg Site data and information index on graph sections (Figure 3.10).

Section	Information gained	From to	No of data points	Information index
1	15	0 to 30	30	0.50
2	7	40 to 160	120	0.06
3	3	170 to 360	190	0.02
4	3.5	380 to 715	335	0.01

The gradient lines indicated that (Figure 3.10; Table 3.1):

- the first line from point 0 to 30 provided an information index[6] of 0.5;
- the second from point 40 to 160 with 0.06;
- third line from point 170 to 360 at 0.02;
- fourth line from point 380 to 715 with 0.01.

The first line has an index of 0.5 and the second line provides a difference in the known information of 0.06, which is almost 10 times less and requires 4 times the number of data points. The last gradient line provides an index of only 0.01 with 335 points which equates to 50 times less information with 11 times more data points. If the analyst considered, e.g., a 10% (or a ±1 m) error as acceptable, then only the first 6 to 10 data points would provide *sufficient* data. Data points in excess of 50 points would only add value to reduce the error margin to below 0.25 m or 3% (Figure 3.6; Figure 3.7).

3.5 DATA-WORTH EVALUATION

The collection of data is associated with cost and time (Figure 3.11). Data therefore have value or are obtained at a cost (Freeze *et al.*, 1992). As data-collection

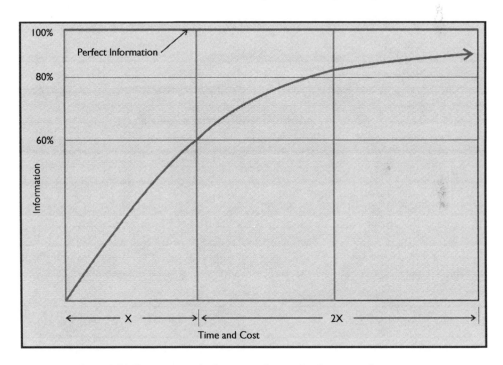

Figure 3.11 Time, cost and information flow in the decision-making process.

6 The information index is defined as the change in the variable with an increasing number of data points, which represents how much we know more against the background of what we already know with each new data point.

programmes could be never ending, data *sufficiency* and *financial* constraints should dictate when data collection would need to end. The concept of *data worth* relates to the value of the data that are collected in site investigations. The first borehole that is drilled or surveyed would, for example, provide the most valuable information in a catchment where there are say no previous boreholes drilled (Figure 3.3). As more boreholes are drilled or surveyed, there would be a point where there are sufficient, say 30 or 50 boreholes surveyed so that new additions would provide fewer data in terms of value than the actual cost of the borehole or the survey (Freeze *et al.*, 1992; Dakins *et al.*, 1995). It is therefore not sensible to aim at collection of all the data in a given catchment for a given project, but rather to *suffice* or *optimise* in terms of statistical representativeness and information provided.

The *optimal* point for data collection would be if the percentage error is below a preset value as determined by the analyst (e.g., <5%) or at the plateau of the logarithmic plot of information flow (Figure 3.8; Figure 3.11). To evaluate the cost of data collection for the Middelburg Site, it was determined that to collect one data point costs R200 (Table 3.2). This cost includes the time of the field surveyor that includes travelling and accommodation. The total cost for surveying 715 data points amounts to R142 800. The information gained is the difference between, for example, the average depth-to-water level with each additional data point gathered (Figure 3.10). The cost per information unit is represented by:

$$CU = \frac{CS}{IG} \tag{3.3}$$

where:
  CU is the cost per unit provided by the information gained (IG)
  CS is the cost per section.

The analysis shows that the cost per unit becomes exponentially more expensive with the number of data points (Table 3.2).

The cost per information unit for line 1, is R400 and for line 4, it is more than R22 000 (Figure 3.10). The information gained and cost can be *optimised* by plotting the information gained and costs against the number of data points (Figure 3.12). The *optimal* number of data points to gain information based on costs is at 149 data points at a cost of R29 600. The gathering of more data points would result in a higher cost relative to the information gained. *Sufficient* information could be

Table 3.2 Middelburg Site data, information and cost of data (Figure 3.12).

Section	No of data points	Information gained	From to	Information Index	Cost per data point (R)	Cost per section (R)	Cost per unit information (R)
1	30	15	0 to 30	0.50	R200	R6 000	R400
2	120	7	40 to 160	0.06	R200	R24 000	R3 429
3	190	4	170 to 360	0.02	R200	R38 000	R9 500
4	335	3	380 to 715	0.01	R200	R67 000	R22 333

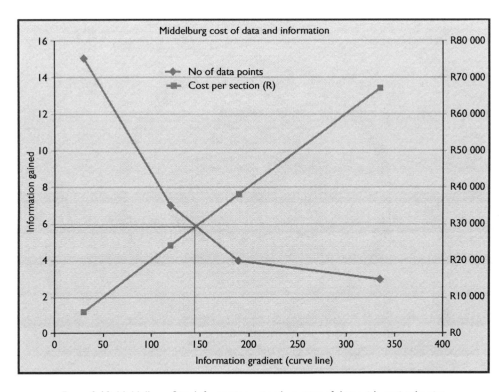

Figure 3.12 Middelburg Site: Information gained vs. cost of data with optimal point.

determined by the analyst at an acceptable error value of e.g., <5%, which is at 30 data points (Figure 3.7; Figure 3.12). The cost of sufficient information after 30 data points is R6 000, which is substantially less than the R29 600 of the cost of the optimal number of data points.

Based on the cost and information considerations, it would be sensible to scale data-collection programmes to start with, aiming to obtain *sufficient* and then *optimal* information. It is therefore not true that 'more is better' during data-collection programmes. From a sustainability and practical perspective, the value of the data vs. the information gained should be considered first.

3.6 INFORMATION AND THE DECISION-MAKING PROCESS

The flow of information determines the nature of the decision-making process. With too little information available the decision-maker would have to make a call based on what is available or take risks on inadequate information. Too much information could lead to unnecessary expenses in terms of time and cost where there are sufficient data pixels to produce an information picture, which then becomes overpopulated when unnecessary data-gathering is continued (Goldratt, 1994). The logarithmic nature of the

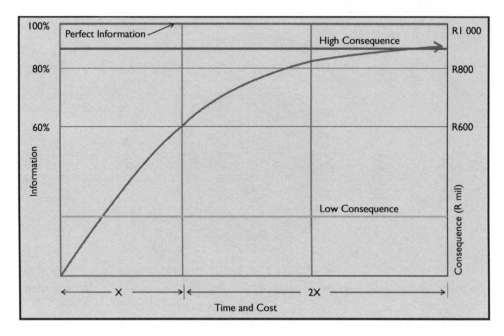

Figure 3.13 Time, cost, information flow and consequence in the decision-making process.

information flow process means that there are points of *sufficient* and *optimal* information to base decisions on and that information decreases exponentially with an increase in data points. A problem that is often encountered in hydrogeological and other environmental investigations is that the analysts either have too few data or aim to obtain *as many data as possible in the pursuit of perfect information*. To aim and obtain perfect information is not possible as it would amount to infinite time and cost. The data-gathering viewpoint is based on the assumption that *more is better*, which leads to the proverbial *analysis paralysis* problem that is usually taken by risk-averse analysts.

The analysis of the decision-making process indicated that the analyst must understand the decision-making process and plan the data-gathering exercise accordingly. If data are not gathered for the purposes of decision-making, then the data-gathering exercise is futile.[7] The decision-making process indicates that it is possible to obtain, for example, 60% of information with a certain amount of time (effort) and cost (equal to X), but to get from 60% to 80% certainty in terms of information, time and cost could be double (+2X) (Figure 3.11). From there it becomes exponentially more difficult to gain additional information and it would take much more to get to, for instance, 80% information, while to get to 100% or perfect information is impossible (Figure 3.11).

Risk is a function of the probability of failure and the cost of the consequence. Risk or consequence is an important consideration that must be included in the decision-making process (Prof JH Venter, Mathematical Statistics, North-West University, Potchefstroom,

---

7 There may be exceptions in research projects, but this study is aimed at practical cases.

Personal Communication, 2010). In high-risk programmes, such as the design of a Nuclear Power Plant (NPP) or a Major Hazardous Installation (MHI), the consequence could become the limiting or determining factor for data-gathering through site investigations and impact assessments. With a higher consequence, the level of information required for decision-making is also higher and vice versa (Figure 3.13). For instance, if the information level is high at 95%, but the cost of failure is say R10 billion and a probability of failure is 0.1%, then the risk is R10 million. In programmes like these, the level of data-gathering should be matched by the level of risk. Consequently if the risk is low, then the level of data-gathering should be lower, in line with the consequence level.

3.7 THE ROLE OF ASSUMPTIONS

In the absence of perfect information, *assumptions*[8] must be made. Negative considerations on assumptions in numerous projects lead to the investigation of the role of assumptions in the decision-making process. The negative view on assumptions is based on an expectation that it can be replaced by (perfect) data. In reality, assumptions are also part of the data-collection and interpretation process (Figure 3.14). For example, the derivation of aquifer parameters is based on analytical and numerical

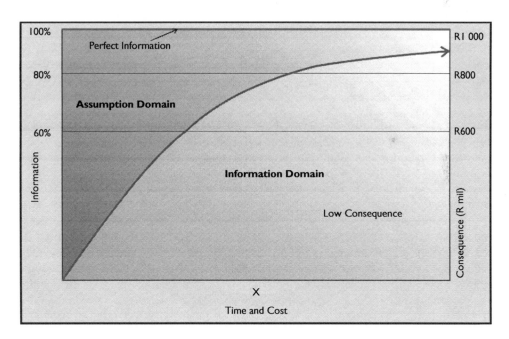

Figure 3.14 Assumption domain and information domain in the decision-making process.

8 A belief or feeling that something is true or will happen, although there is no proof (Oxford English Dictionary 2006). A statement that is used as a premise of a particular argument that may not be otherwise accepted (Collins English Dictionary 2006).

analysis techniques or models that make assumptions in order to arrive at solutions. The well-known Theis or Cooper-Jacob assumptions are applied to aquifers that are assumed to be two-dimensional, and assume horizontal flow, uniform thickness, and homogeneous aquifers of infinite extent (Kruseman and De Ridder, 1991). Except for the three directly measurable data parameters in groundwater (water levels, abstraction rates and water quality), assumptions form an integral part of the data-interpretation and decision-making process in hydrogeology.

In a decision-making process, assumptions are used to substitute information. Assumptions could be used because of limited time and budget constraints to obtain the information or when it is not possible to make a conclusive argument based on the information. In many cases, some of the information is impossible to obtain. For example, it is not possible to determine exactly all the underground fracture zones and preferential pathways or recharge in an aquifer. Assumptions are replaced by information as the decision-making process develops by spending more time and money (Figure 3.14).

It is proposed that in the absence of scientific information, *conservative assumptions* should be used, which is in line with the *precautionary principle* (National Environmental Management Act; NEMA, No 107 of 1998). Assumptions used in this way would always have the effect that, for instance, more water is available for environmental use than the assumed case. In the case of underground mine-flooding evaluations, the assumption would be to determine the upper bound of inflow. Hence to be conservative, assumptions may in the same investigation seem to be contradictory.

3.8   CONCLUSIONS

The role of data and information was investigated to determine when data are enough for the purposes of decision-making. The aim was to characterise the decision process based on the influence of data and information on it. The decision-making process is based on the availability of information. Data are gathered to provide information which when analysed becomes knowledge that is used to make management decisions.

To determine the effect of data and information on the decision-making process, data from a field site were evaluated. The depth-to-groundwater level was used as the required variable on which information was required. The change in the average depth-to-groundwater level was evaluated against the increase in the number of data points. Information was gained based on the change that a data point provides against the previous value. If the average depth-to-groundwater level after e.g., 3 data points is 8 m and after e.g., the fifth data point is 10 m, then the fifth data point provided information as there was a change. This process was done for all 715 data points collected for the Middelburg Site. It indicated that the information process is logarithmic. The percentage error in the average becomes small at <5% after 30 points and very small at 0.6% after 100 data points and eventually 0.08% after 715 data points. Although it approaches zero, it is asymptotic and will never reach it. The concept of perfect data (does not exist, but was used in the same way that an ideal gas is defined in physics) is defined as data with a standard deviation of zero. In the case of perfect data, perfect information can be obtained by using one data point. Perfect information would therefore amount to 100% certainty and would idealistically allow the decision-maker to make the perfect decision.

The analysis showed that the decision-making process has a logarithmic nature which means that less and less information becomes available with more data. It is similar to the law of diminishing returns that is used in economics. *More data do not necessarily mean more information* and are not necessarily better for decision-making. The value of data was assessed in a data-worth or data-cost evaluation. The information gained and cost can be optimised by evaluating it against the number of data points. The optimal number of data points to gain information based on costs for the Middelburg Site is at 149 data points and at a cost of R29 600. The gathering of more data points would result in a higher cost relative to the information gained. *Sufficient* information could be determined by the analyst at an error value of e.g., <5%, which is at a lower number of data points, which was 30 in the case study. The cost of the sufficient information after 30 data points is R6 000, which is substantially less than the R29 600 of the cost of the *optimal* number of data points. More data are also not better from a sustainability perspective that includes the impact on cost.

First of all, a point of *sufficient* information should be determined by the analyst, which, if justified economically, can be increased to *optimal* information in an iterative process. The effect of risk or consequence is important in the information and decision-making process. With a higher consequence, the level of information required for decision-making is also higher and vice versa. Assumptions form an integral part of the data-interpretation and decision-making process. In a decision-making process, assumptions are used to substitute information. It is used, because there is not sufficient time and budget available to obtain the information or it is impossible to make a conclusive argument based on the information at hand. It is proposed that conservative assumptions be used, in line with the precautionary principle for the purposes of decision-making in environmental groundwater-management problems.

The outcomes from this study indicate that the data-gathering programmes should be planned to be *iterative* in cases where the cost of data gathering would be high. The data worth could be determined either by the sufficiency of the initial data-gathering exercise to determine how much time and effort should be allocated to for example additional data-gathering exercises.

3.9 RECOMMENDATIONS

Based on the findings of this study, the following recommendations are made:

- Statistical methods should be used within the constraints of the objective of the project to plan data-gathering exercises;
- Groundwater specialists should be trained in basic decision-making principles and processes;
- Additional research should be done to determine the characteristics that influence the information gained and the slope of the information curves that were obtained in the field study.

REFERENCES

AGES (2009) *Middelburg groundwater investigation*. AGES internal report.

AGES (2010c) *Middelburg groundwater investigation geophysics, exploration drilling and aquifer testing phase report M2*. Pretoria, Gauteng, South Africa, AGES Consulting. AGES report no: 2009/01/05/GWSE.

AGES (2010d) *Groundwater specialist investigation at a platinum mine in the North-West Province of South Africa*. Pretoria, Gauteng, South Africa, AGES Consulting. AGES internal report no: AS-R-2010-06-02.

AGES (2010e) *Groundwater specialist investigation at Zandrivierspoort north of Polokwane*. Pretoria, Gauteng, South Africa, AGES Consulting. AGES internal report no: AS-R-2010-12-10.

Beuche, F.J. (1986) *Introduction to Physics for Scientists and Engineers*. New York, NY, McGraw-Hill Book Company.

Collins English Dictionary and Thesaurus (2006) Updated 4th edition. Glasgow, UK, Harper Collins Publishers.

Dakins, M.E., Toil, J.E., Small, M.J. & Brand, K.P. (1995) *Risk-Based Environmental Remediation: Bayesian Monte Carlo Analysis and the Expected Value of Sample Information*. [Online] Available from: http://onlinelibrary.wiley.com [Accessed July 2011].

Freeze, A.R., James, B., Massman, J., Sperling, T. & Smith, L. (1992) Hydrogeological decision analysis 4. The concept of data worth and its use in the development of site investigation strategies. *Groundwater*, 30, pp. 574–588.

Goldratt, E.M. (1994) *The Haystack Syndrome: Sifting Information Out of the Data Ocean*. New York, NY, North River Press.

Handley, J.R.F. (2004) *Historic Overview of the Witwatersrand Goldfields*. Howick, South Africa, Author.

Kruseman, G.P. & De Ridder, N.A. (1991) *Analysis and Evaluation of Pumping Test Data*. 2nd edition. Wageningen, The Netherlands, International Institute for Land Reclamation and Improvement.

Mohr, P. & Fourie, L. (2004) *Economics for South African Students*. 3rd edition. Paarl, South Africa, Van Schaik.

NEMA (1998) National Environmental Management Act (NEMA, No 107 of 1998). *Government Gazette* Vol. 401, No 19519. 27 November 1998. Cape Town, South Africa.

Oxford Advanced Learners Dictionary of Current English (2006). 7th edition. Oxford, UK, Oxford University Press.

Venter, J.H. (2010) Professor J.H. Venter, Department of Mathematical Statistics, North-West University, Potchefstroom, South Africa. (Personal communication, July 2011).

Chapter 4

Development of emergency water supplies for the drought-impacted southern Cape coastal region of South Africa – observations while abstracting saline water for desalination

Roger Parsons
Parsons & Associates Specialist Groundwater Consultants, South Africa

ABSTRACT

While implementing emergency water-supply schemes for the drought-impacted southern Cape coastal region of South Africa, boreholes were drilled directly adjacent to the Knysna Estuary to provide saline feed water to the desalination plant. The siting of the boreholes was based on a conceptual understanding of the position of the saline water – freshwater interface provided by the Ghyben-Herzberg relationship. The water yielded by the boreholes during testing and production was of lower salinity than expected. It was found that low-salinity groundwater initially provided about half the water to the boreholes, gradually decreasing to 28% over a period of 7 months of continuous pumping. The salinity gradually increased from 2 600 mS/m to 4 000 mS/m as the relative contribution of seawater from the estuary increased. These observations highlighted the fact that the Ghyben-Herzberg relationship is based only on the density of two fluids, and it fails to recognise the hydrodynamics of the subsurface environment, particularly during periods of groundwater abstraction.

4.1 INTRODUCTION

The onset of drought in the southern Cape coastal region of South Africa placed water supplies in the region under considerable pressure, with rainfall reported to be the lowest measured in the past 130 years. Impacts of the drought first manifested themselves in Sedgefield, a small coastal holiday village, in January 2009. Sedgefield obtains its water supply from a run-of-river abstraction point in the Karatara River. When flow in the river ceased, water had to be trucked in from George located some 50 km to the west. Intermittent flow in the Gouna River and low dam levels threatened the water supply to Knysna as well as the tourism industry on which the economy of the town is highly dependent. The area was declared a disaster area in November 2009, and urgent intervention required that a 6 Ml/d emergency water supply be developed. Development of groundwater supplies, seawater desalination, recycling of wastewater and establishment of a regional pipeline were some of the options considered. In the case of Knysna it was decided to develop a 2 Ml/d desalination plant in the town and source a further 2 Ml/d from groundwater. The balance of the water supply was to be sourced from existing supplies which still yielded small amounts of water. This paper documents experiences and observations during the installation and testing of boreholes used to provide source water to the desalination plant.

4.2 CONCEPTUAL MODEL

Saltwater intrusion as a result of groundwater abstraction is a well-recognised problem, and as a result the Ghyben-Herzberg relationship (Davis and De Wiest, 1966) is included in basic hydrogeological teaching. The depth of the interface between immiscible freshwater and saltwater bodies in contact with each other is controlled by the density of the two fluids, with the relationship being expressed as follows (Figure 4.1):

$$Z_s = \frac{P_f Z_w}{P_s - P_f}$$

where:

Z_s = depth of interface below mean sea level
Z_w = height of water table above mean sea level
P_f = density of freshwater
P_s = density of saltwater

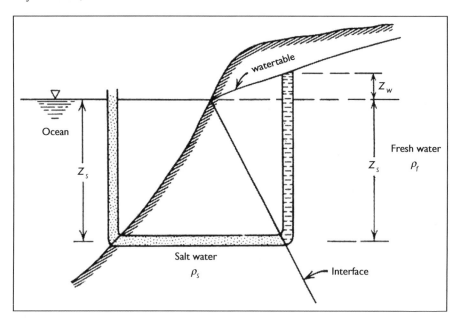

Figure 4.1 Position of the saline water – freshwater interface according to the Ghyben-Herzberg relationship (from Davis and De Wiest, 1966).

Accepting P_f and P_s to be 1.000 g/cm^3 and 1.026 g/cm^3, the relationship is simplified as:

$$Z_s = 38 \cdot Z_w$$

Hubbert (1940) recognised that the assumption that groundwater and seawater were in a state of hydrostatic equilibrium was invalid, as groundwater is in a state of constant motion. The position of the saline water – freshwater interface would be seawards of that predicted by the Ghyben-Herzberg relationship. Kohout (1960) reported this to be the case in his seminal study of the Bay of Biscayne in south-eastern Florida. Field measurement confirmed the interface not to be a sharp transition, but diffuse in character and in excess of 100 m wide.

In spite of its simplicity, the theory is widely considered in the management of coastal aquifers. While saltwater intrusion is not a major problem in South Africa, isolated and local-scale instances have been reported. Because of the threat of saltwater intrusion, South African water legislation stipulates that no groundwater abstraction is generally authorised within 750 m of the coast and any abstraction in excess of Schedule 1 Use (essentially limited to small-scale domestic use and stock watering) is subject to a water-use licence being issued by the Department of Water Affairs.

Based on their understanding of the Ghyben-Herzberg relationship, hydrogeologists usually aim to prevent saltwater intrusion by limiting rates of abstraction (and hence limiting drawdown) and proximity to the coast. In the case of establishing a source of seawater for the Knysna desalination plant, it was hypothesised that the reverse relationship could be applied, i.e. boreholes were to be drilled as close to the seawater as practically possible and high pumping rates were to be adopted so that seawater could be induced into boreholes and thus provide the constant quality required for the desalination process.

4.3   STUDY SITE

Knysna is located in the Western Cape Province of South Africa between Cape Town and Port Elizabeth (Figure 4.2). The town has a permanent population of about 70 000. It was established on the eastern banks of the Knysna Estuary with the amalgamation of two hamlets in 1882 that served the forestry and shipping industries. The area experiences a temperate coastal climate with mean annual precipitation amounting to 700 mm/a at the coast. Rain falls throughout the year.

It was decided to establish the desalination plant in the vicinity of Loerie Park as the land was municipal-owned, sufficient space existed on which to establish the plant, the site was directly adjacent to seawater and the produced water could easily be added to the town's water reticulation scheme. After some preliminary investigations, it was also decided to abstract the source water from a series of boreholes drilled on the edge of the Knysna Estuary, rather than use a more expensive direct off-take system that provided environmental challenges from the sensitive estuary. Boreholes were drilled at the high-tide mark on the eastern banks of the Ashmead Channel between Thesen Island and the mainland. The site is flat-lying and located directly west of the Knysna wastewater treatment works which has a treatment capacity of 6.7 Ml/d.

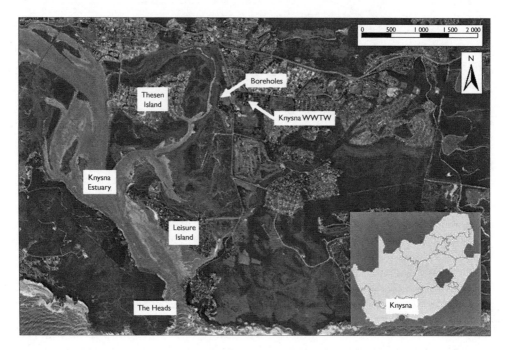

Figure 4.2 Locality map of the boreholes used to provide feed water to the Knysna desalination plant.

The Knysna Estuary is classified as an estuarine bay or marine embayment (Grindley, 1985); and is considered the most important of South Africa's 258 estuaries in terms of conservation importance (Turpie *et al.*, 2004). Water in the estuary is saline, with the tidal reach extending some 19 km up river. The study site is in the bay regime of the estuary which exhibits salinities similar to the ocean (~5 500 mS/m) (Largier *et al.*, 2000 as cited in Russell *et al.*, 2009).

The study area is underlain by folded, fractured and faulted quartzite and quartzitic sandstones of the Silurian-aged Table Mountain Group that form part of the regionally extensive and significant Table Mountain Group aquifer system. Locally, these are overlain by unconsolidated fine-grained sand of varying thickness that forms a local primary aquifer system (Johnstone, 2009). Depending on the location in the estuary, the sand can be of marine, fluvial or aeolian origin. The deposits underlying the site were interpreted to be of fluvial origin.

4.4 BOREHOLE ESTABLISHMENT

Installing and pumping two well points on the water's edge showed that this method of abstraction was not suitable for providing feed water to the plant. In late November 2009 a pilot borehole was then drilled using a rotary percussion drilling rig to assess the nature of the subsurface and the hydrogeological characteristics. Unconsolidated fine-grained sand was encountered to a depth of 15 m, whereafter drilling proceeded through quartzitic pebbles and boulders to a depth of 25 m. Un-perforated 165 mm steel casing was installed to a depth of 17 m, but continual collapse of the borehole prevented deeper installation and proper completion of the borehole. In light of the high blow yield (8 l/s) it was decided to test the borehole. A step-drawdown test comprising four steps ranging between 3.3 l/s and 5.7 l/s was conducted, with the narrow diameter limiting the capacity of the test pump that could be used. Pumping induced a drawdown of about 12 m and water with an Electrical Conductivity (EC) of 5 400 mS/m, i.e. almost seawater.

On the basis of the results from the pilot borehole, it was decided to drill six boreholes to provide the 4 Ml/d required for the desalination plant. The ODEX drilling method had to be used to counter the collapsing pebbles and boulders. A special off-centre percussion hammer is used that allows the drilling of a hole bigger than the casing. The casing is attached to the drill stem by means of a shoe and is installed while drilling. On reaching the required depth, the eccentric bit is centred and extracted inside the casing. The inner diameter of the 200 mm ODEX shoe typically used in the area limited the final inner diameter of the borehole to 165 mm. The cost of importing a larger shoe and the time required to get the equipment onto site was considered prohibitive in the context of the urgency of the project.

Nine boreholes were drilled at the site during January 2010 and February 2010, with a revised target of 6 Ml/d. Three additional boreholes were drilled in February 2011. The boreholes were numbered using the prefix LPSWE, i.e. Loerie Park Seawater Extraction (Figure 4.3). The general profile of the boreholes comprised 14 m of fine sand overlying 6 m of well-rounded but poorly sorted pebble and boulder horizon, which in turn overlay a greenish marine clay. Drilling was stopped when the clay

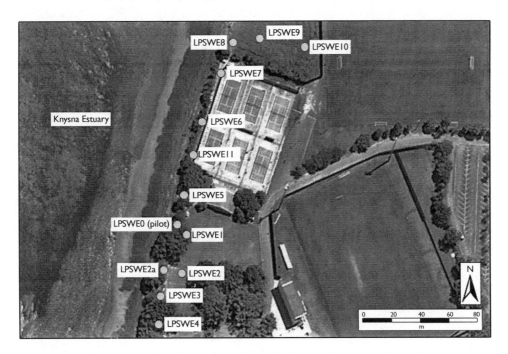

Figure 4.3 Position of the boreholes at Loerie Park in relation to the Knysna Estuary.

was reached. The boreholes were fitted with 165 mm to 144 mm PVC casing, with the bottom third of the borehole screened with 0.5 mm slots. The annulus was backfilled with 16/30 filter sand and the borehole was air-flushed for 1 h on completion.

All boreholes were subject to step-drawdown tests ranging between four and eight steps to develop the boreholes, determine the yield, and assess the quality of the water abstracted. LPSWE2 was subjected to a 24 h constant discharge test. In addition to salinity issues, the groundwater had a high colour and turbidity which posed problems with the water-treatment process.

4.5 RESULTS AND OBSERVATIONS

Excluding LPSWE2a (low yielding and too close to LPSWE2) and LPSWE10 (very low salinity because too far from the water's edge), the remaining ten boreholes yielded 83.5 l/s (or 7.2 Ml/d), with an 'average' EC of almost 3 000 mS/m (Table 4.1). While borehole yields were generally in line with expectations, the salinity of the abstracted water was surprisingly low. Water in the estuary at site was measured to have an EC of 5 500 mS/m, but some of the boreholes yielded half that.

It was also observed that the salinity often improved during the pumping tests, rather than deteriorate as expected from the conceptual model. By way of example, monitored EC levels during the testing of LPSWE5 and LPSWE8 are presented in Figure 4.4 and Figure 4.5. In the case of LPSWE5, EC started at 4 000 mS/m and dropped to 2 000 mS/m by the end of the first step. Thereafter, the EC increased to 5 000 mS/m by the end of the third step and then gradually decreased to 4 100 mS/m over the next 3 h of pumping. On cessation of pumping, the EC rose to almost 5 700 mS/m. By contrast, the EC monitored at LPSWE8 remained relatively constant, gradually decreasing from 2 600 mS/m to 2 300 mS/m over the duration of the 5 h step-drawdown test.

Table 4.1 Recommended borehole yields and water quality of boreholes drilled to provide feed water to the Knysna desalination plant.

Borehole No.	Borehole depth (m)	EC (mS/m)	Pump install depth (m)	Yield (l/s)
LPSWE1	25	3200	24	10.0
LPSWE2	24	1600	22	10.0
LPSWE2a	21	4500	20	(4)
LPSWE3	20	2600	18	14.0
LPSWE4	13	1800	12	6.5
LPSWE5	22	5700	21	7.5
LPSWE6	20	2300	19	10.0
LPSWE7	19	3000	18	8.0
LPSWE8	15	2300	14	7.5
LPSWE9	15	2200	14	5.0
LPSWE10	14	350	13	8.0
LPSWE11	18	5000	17	10.0

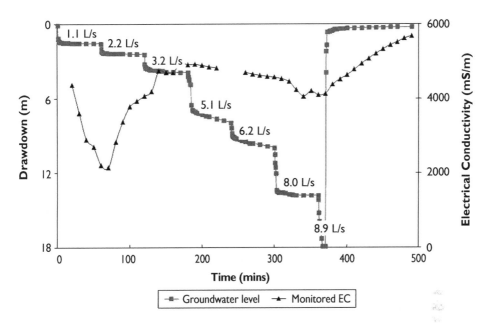

Figure 4.4 Salinity variations monitored during the step-drawdown pumping test conducted on LPSWE5.

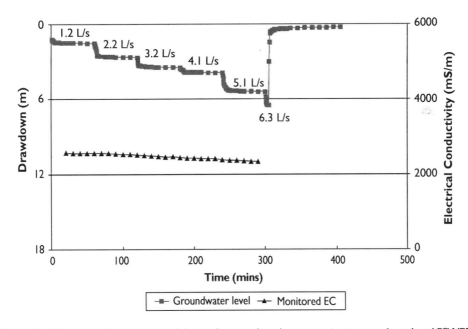

Figure 4.5 Salinity variations monitored during the step-drawdown pumping test conducted on LPSWE8.

When the desalination plant was first brought into production in June 2010, eight boreholes were collectively pumped almost continuously at 60 l/s (or 5.2 Ml/d). The ratio of feed water to product water was 2.6:1. Between June 2010 and January 2011, the 'average' EC deteriorated from 2 600 mS/m to 4 000 mS/m (Figure 4.6).

4.6   DISCUSSION

EC data monitored during both testing and production indicates that some freshwater is being abstracted from the boreholes, as the abstracted water had an EC intermediate between that of seawater (5 500 mS/m) and that typical of groundwater in the vicinity (100 mS/m). Using these two values as end members of a mixing model, it was calculated that at the start of production in June 2010 seawater and groundwater contributed a similar volume of water abstracted from the boreholes. However, the seawater contribution gradually increased with continued pumping and by January 2011 almost 72% of the water abstracted was from the estuary. The primary aquifer contributed only 28% of the water abstracted.

With the positioning of the boreholes at the interface between seawater and groundwater at ground level and adopting of high pumping rates, it was hypothesised that the boreholes would yield seawater. However, the hypothesis proved to be false, as demonstrated by the observed freshwater contribution from the transmissive primary aquifer. The conceptual model presented in Figure 4.1 is used primarily to explain the position of the interface based on the difference in fluid densities. It is also used as a basis for developing groundwater-abstraction strategies to prevent saltwater intrusion.

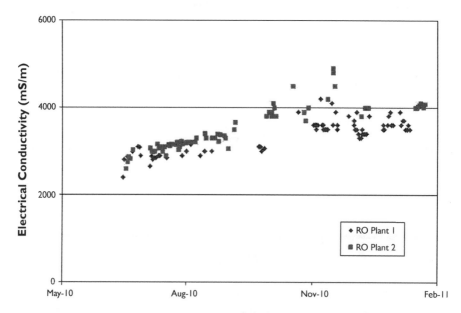

Figure 4.6 Salinity variations monitored during the commissioning and testing of the Knysna desalination plant between June 2010 and January 2011.

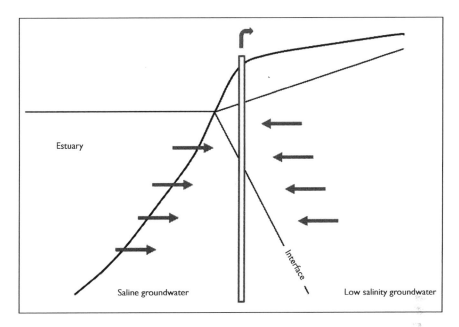

Figure 4.7 Modified conceptual model taking into account flow dynamics during pumping.

However, the reverse application of the conceptual model failed to appreciate the landward groundwater contribution to boreholes drilled at the interface. Initially, the contribution from the landward side was equal to that from the estuary (Figure 4.7). While the head in the estuary remained nearly constant (save for tidal fluctuations), the groundwater level on the landward side was lowered by between 6 m and 12 m in response to abstraction. In turn, this resulted in a flattening of the hydraulic gradient on the landward side and a reduction in contribution to the production boreholes as observed during continuous pumping.

The use of boreholes to abstract feed water for desalination plants remains a hydrogeological matter, but it does require a different conceptualisation of the situation. While the Ghyben-Herzberg relationship – with its inherent shortcomings – can be used to understand the position of the interface, it does not conceptualise flow dynamics. The modified conceptual model presented in Figure 4.7 can be used to better explain the quality of water abstracted from boreholes at the saline water – freshwater interface during pumping.

4.7   CONCLUSIONS

The Ghyben-Herzberg relationship is widely used to explain and understand the position of the interface between seawater and groundwater. The relationship is based only on the density of the two fluids, and fails to take cognisance of flow dynamics and flow boundaries. Failure to appreciate the contribution from groundwater to boreholes drilled directly adjacent to the Knysna Estuary resulted in the boreholes yielding water of a lower salinity than anticipated. Continual pumping over a 7-month period resulted in the groundwater contribution decreasing from 50% to 28%, with a concomitant increase in salinity from 2 600 mS/m to 4 000 mS/m. Appreciation of the modified conceptual model that takes into account flow dynamics during pumping can prevent similar misinterpretations in the future.

ACKNOWLEDGEMENTS

The support of the Knysna Municipality and project engineers SSI Consulting Engineers is acknowledged, in particular Messrs Rhoydan Parry, Hennie Erwee and Keith Turner. The efforts of Gary Price (Gary's Borehole Solutions) and Edwin Gerber (AB Pumps) and their teams during the project are appreciated, as are the constructive comments of the paper reviewer.

REFERENCES

Davis, S.N. & de Wiest, R.J.M. (1966) *Hydrogeology.* New York, NY, John Wiley & Sons.

Grindley, J.R. (1985) *Estuaries of the Cape – Part II – Synopses of available information on individual systems.* Stellenbosch, South Africa, Council for Scientific and Industrial Research. Report No. 30 Knysna (CMS 13), CSIR Research Report 429.

Hubbert, M.K. (1940) The theory of groundwater motion. *Journal of Geology* 48, 785–944.

Johnstone, A. (2009) *Hydrogeology of Leisure Isle, Knysna.* Johannesburg, South Africa, GCS (Pty) Ltd. Report LIRA.07.142.

Kohout, F.A. (1960) Cyclic flow of salt water in the Biscayne aquifer of southeastern Florida. *Journal of Geophysics Research* 65, 2133–2141.

Russell, I.A., Randall, R.M. & Kruger, N. (2009) *Garden Route National Park Knysna Coastal Section – State of Knowledge.* Sedgefield, South Africa, SANParks Scientific Services.

Turpie, J.K., Adams, J.B., Joubert, A., Harrison, T.D., Colloty, B.M., Maree, R.C., Whitfield, A.K., Wooldridge, T.H., Lamberth, S.J., Taljaard, S. & Van Niekerk, L. (2004) Assessment of the conservation priority status of South African estuaries for use in management and water allocation. *Water SA* 28, 191–206.

Chapter 5

Guidelines for integrated catchment monitoring: ICM mind-map development and example of application

N. Jovanovic[1], S. Israel[1], C. Petersen[1], R.D.H. Bugan[1], G. Tredoux[1], W.P. de Clercq[2], R. Rose[3], J. Conrad[3] & M. Demlie[4]

[1]CSIR, Natural Resources and Environment, Stellenbosch, South Africa
[2]Department of Soil Science, University of Stellenbosch, South Africa
[3]GEOSS – Geohydrological & Spatial Solutions International (Pty) Ltd., Stellenbosch, South Africa
[4]Department of Geology, University of KwaZulu-Natal, Durban, South Africa

ABSTRACT

Advances have been made in recent years in developing networks and databases for monitoring water systems in South Africa. In this study, an Integrated Catchment Monitoring (ICM) mind map was developed to include guidelines and recommendations on the minimum monitoring requirements (e.g., site selection, type of variables, space and time frequency, methodologies, data handling and quality assurance, input data requirements for models), as well as essential information and sources (e.g., available databases, roles and responsibilities in monitoring). The user-friendly ICM mind map was written in the open source software FreeMind v. 0.9.0. The application of the ICM mind map and the benefits of comprehensive monitoring were demonstrated in the pilot catchment of the Sandspruit River, a seasonal tributary of the Berg River (South Africa). The ICM mind map was used to assess monitoring gaps, to expand the monitoring network (weather monitoring, hydrometry, vadose-zone profiling, geophysical and isotope studies) and to quantify the water balance. The combination of monitoring and modelling of all water-cycle components proved to be a powerful tool. The main target users of the ICM mind map are water management boards and similar institutions.

5.1 INTRODUCTION

Advances have been made in recent years in developing networks and databases for monitoring water systems with the ultimate aim of facilitating integrated water-resource management at a catchment scale. These include guidelines and networks for monitoring groundwater and rainfall in South Africa (DWAF-DANCED 2007a and b), as well as for surface water (UNEP/WHO, 1994; Oakley *et al.*, 2003; Dallas and Day, 2004; Strobl *et al.*, 2006). However, to facilitate integrated catchment monitoring (ICM), these monitoring systems need to be consolidated and integrated amongst various components of catchment systems, i.e., groundwater, surface water, soil and vadose zone (unsaturated zone, including surface water – groundwater interactions) and atmospheric monitoring (including rainfall and evapotranspiration). Each of these components requires collection and management of purposeful and relevant data to address the main problems identified in current hydrological research and practice, e.g. groundwater protection, groundwater recharge and groundwater – surface water interactions. Data collection and management is currently done by different government departments, institutions, firms, etc. for each component of the catchment system. Given the interactions between these components, it was essential to develop guidelines for coordination of data collection, management and exchange amongst the different stakeholders (e.g. defining roles and responsibilities for data collection and management).

Monitoring programmes need to be reliable and make use of state-of-the-art monitoring technologies, while collected data need to be easy to handle and compatible with models. In addition, monitoring is often limited in time (e.g. during individual research projects), space (e.g. areas of ecological sensitivity or hotspots) or type (e.g. specific contaminants) and this needs to be addressed in order to assist management of water resources at catchment level. Another important variable to account for in the monitoring framework is different land uses (Usher *et al.*, 2004). Land use and associated sources of pollution determine the intensity and type of monitoring to be done. Many authors (Chapman, 1992; Canter *et al.*, 1987; Nacht, 1983) have emphasised the need to define clearly the objective of a groundwater monitoring programme before beginning any monitoring design so as to get adequate information. Real-time data collection with sensors is favoured, but it is expensive, hence the need for optimising the use and quality of data collection. Appropriate scales and frequencies of data collection and management have to be defined in order to facilitate the integration of the different components of catchment systems. Optimisation of monitoring networks has to account for both spatial and temporal variations through the application of geostatistical analyses, as well as logistical and financial feasibility.

As water managers are usually not experts in all relevant disciplines (hydrogeology, hydrology, soil science, meteorology), there was a need to provide a product that would direct the user/water manager towards the appropriate guidelines, database, methodology or information. The key question in this study was: 'What does a catchment manager need?' The main objective of this study was the development of guidelines for monitoring best practices applicable to South(ern) African conditions, for the different components of catchment systems (groundwater, surface water, soil and vadose zone, atmosphere). The development of a user-friendly tool incorporating guidelines for ICM (ICM mind map) and its application are presented.

5.2 DEVELOPMENT OF THE ICM MIND MAP

5.2.1 Software development

The ICM mind map is a user-friendly tool that enables water managers to find and access any information and guidelines for monitoring any of the components of the environment (groundwater, surface water, soil and vadose zone, atmosphere, river health) related to the catchment water cycle. This tool indicates the minimum monitoring requirements (e.g. type of variables, space and time frequency), and provides essential information and sources of information in order to obtain a meaningful amount of data for a specific monitoring objective. Some of the principles used to develop the ICM mind map include the requirement for practical, user-friendly and accessible guidelines. For this purpose, a freely downloadable mind-mapping software, called FreeMind v. 0.9.0, was used. The content of the ICM mind map was compiled using existing and newly developed guidelines as well as through several stakeholder workshops and meetings. The content was described in detail by Jovanovic *et al.* (2011). Examples of existing guidelines can be found in DWAF (1997), DWAF (2004a and b), DWAF (2006) and DWAF (2007). The ICM program can be easily expanded and updated by including any additional information.

A sketch of the full tree of the ICM mind map, compiled in FreeMind, is shown in Figure 5.1. The primary branches of the ICM mind map are: (i) groundwater monitoring; (ii) hydrological monitoring; (iii) atmospheric monitoring; (iv) soil and vadose zone; (v) river health; (vi) matrix of monitoring objectives vs. space and time frequency of monitoring variables; and (vii) modelling input data requirements (summary of typical minimum requirements for input data in hydrological and groundwater models). Secondary branches can be accessed from the primary branches and so on (Figure 5.1). Each component of the environment includes the following items: objectives of monitoring and applications; users of monitoring; available databases, roles and responsibilities in monitoring; type of monitoring variables; selection of monitoring sites, space and time frequency of monitoring; methodologies in monitoring (sampling, analytical procedures, data capture, handling, presentation of results and quality assurance), and inventory of hardware and accredited laboratories country-wide. Each of these items includes hyperlinks. The hyperlinks are used to connect directly to guideline documents (Word or pdf files) and/or web sites (e.g. database of South African Weather Services for atmospheric monitoring).

The ICM mind map makes provision for data exchange, actions and interactions between custodians in data collection, capture, storage and management, through the inclusion of specific functions (Figure 5.1): (i) interactions (e.g. monitoring of water flow and quality concurrently); (ii) data exchange (e.g. how to exchange rainfall records between South African Weather Services and Department of Water Affairs); and (iii) action points (e.g. collecting soil-survey data from individual studies and collate them into a central database managed by the Provincial Departments of Agriculture). The matrix of monitoring objectives vs. space and time frequency of monitoring variables includes a summary of monitoring objectives vs. type of variable, spatial requirements and time frequency for the different components of catchment systems.

5.2.2   Guiding principles

The driving principle of the ICM mind map is monitoring of all water-cycle components, as well as the consequential cause-and-effect relationship between geology, climate, soil and land use. In practice, the recommended catchment-monitoring approach should include the following:

–   hydrometry (surface water and groundwater flow and quality measurements);
–   geophysical studies (analysing the geological layering and water occurrences in the subsoil);

Figure 5.1 Printout of the integrated catchment monitoring mind map compiled in FreeMind v. 0.9.0.

- isotope studies (analysing the flow paths in the catchment and hydrograph separation);
- weather monitoring (representing the driving force of water fluxes in the catchment);
- vadose-zone profiling (analysing the modalities of water flow and contaminant transport in the sub-surface).

During the development of the ICM mind map, knowledge was gained on the interactions between institutions responsible for monitoring and custodians of databases, as well as on sustainable strategies for future monitoring and handling of data. Data and databases should be kept separately for different environmental components and handled by specialists in the particular discipline, whilst data and metadata should be integrated. Modalities of data exchanges between institutions that monitor and manage databases were proposed (e.g. standardised data collection and quality control of rainfall). Data quality control and the importance of using accredited laboratories was stressed in terms of quality assurance (e.g. for chemical analyses). It was recommended to collate existing weather data from research projects, farmers, schools, etc., and existing soil data from universities, municipalities, consulting firms and accredited laboratories.

The catchment scale is the ideal and most appropriate operational scale for management and monitoring of natural resources, water in particular. If South Africa is used as an example, catchment management agencies (CMAs) are responsible for the management of 10 to 20 quaternary catchments per water management area (WMA). Interactions between water-governance levels and three different scales of data collection with a common database are proposed, namely:

- the local data network, having the highest density, should be handled by local government with the objective of day-to-day water supply to users;
- a medium-density network of data should be handled by regional offices of the Department of Water Affairs (DWA) and/or CMAs with the objective of catchment management;
- the national data network, having the lowest density, should be handled by South Africa's DWA Head Office for assessment, management and planning purposes.

Financial limitations are often the biggest constraints in a monitoring programme as they imply practical difficulties in implementing monitoring. These include limited budgets and budget cuts in the government, volatility of yearly government budgets allocated to monitoring, lack of staff and capacity (e.g. in automated recording and real-time data handling), theft and vandalism (e.g. selected sites need to be appropriate but also safe), specific requirements (e.g. an environmental impact assessment is required for the establishment of gauging weirs for surface-flow monitoring), reliability, maintenance and replacement of equipment (e.g. common problems encountered in the field are fires, along with the limited lifespan of equipment). For these reasons, the terminology 'ideal' monitoring network was used as opposed to 'optimal' monitoring network, as it is not always financially feasible to achieve optimal monitoring.

A large number of sampling and training manuals were developed in the past and they reside in government offices. However, these are often not used or they are not found due to turnover of staff, where newly appointed staff members are not always aware of the existence of these documents. Awareness on currently available guidelines and manuals should be increased and an assessment undertaken on the usage and usability of such documents. The ICM mind map facilitates, to a certain extent, water management as all key documents (guidelines, manuals) or links thereto are included in one programme.

5.3 ICM MIND-MAP APPLICATION

Before undertaking a monitoring programme in any region of interest, an initial desk study and review of existing data are prerequisites (topography, climate, geology, hydrology, soils and land use) to serve as baseline data. The Sandspruit catchment, a tributary of the Berg River (Western Cape, South Africa), was used as a pilot study site for the ICM mind-map application. The objective of the monitoring programme was to enable a comprehensive description of the study catchment. The baseline study provided an indication of available data and hence the existence of monitoring gaps. A new expanded monitoring programme was then designed based on the gaps identified and the benefits of this new monitoring programme are illustrated. Throughout the exercise, the principles, guidelines and information included in the ICM mind map were used.

5.3.1 Description of catchment and baseline data collection

The Sandspruit catchment (DWA quaternary catchment G10J) is a seasonal tributary of the Berg River (Western Cape, South Africa), flowing predominantly between May and November. The Sandspruit catchment is approximately 152 km² in size and it is located in the middle reaches of the Berg River basin (Figure 5.2).

The terrain types are low mountains and open high hills or ridges in the upper southern reach of the Sandspruit catchment, rolling and irregular plains with some relief in the middle reaches, and plains with some relief in the lower reaches. The elevation ranges between 40 m above mean sea level (amsl) in the low elevation areas (north-west) to 900 m amsl in the high-elevation southerly parts of the catchment (Table Mountain Group sandstone Kasteelberg). The Berg River flows north-westwards and lies north-east of the catchment. The Berg River catchment experiences a Mediterranean climate with warm dry summers and cool wet winters. Rainfall is generally in the form of frontal rain approaching from the north-west, extending normally over a few days with significant periods of clear weather in between. Mean annual rainfall in the Sandspruit catchment is between 300 mm and 500 mm, mainly in winter (from April to October). The average long-term annual temperature is between 15°C and 21°C. Annual evaporation is predominantly between 1 800 mm and 2 000 mm.

The geology in the Sandspruit catchment shows minimal variation, being dominated by Table Mountain Group (TMG) sandstone in the high-elevation areas and Malmesbury shale in the mid- to low-elevation parts. An alluvium cover is also evident, which increases in thickness towards the lower elevation areas of the catchment. The Malmesbury Group rocks form low to moderately productive fractured aquifers. Although these aquifers

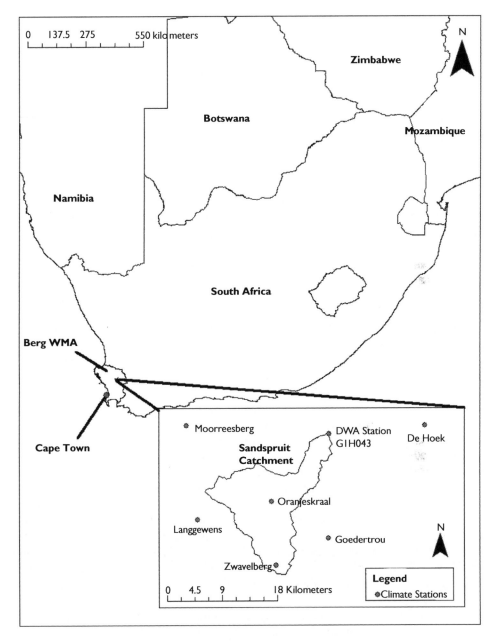

Figure 5.2 Location of Sandspruit catchment within the Berg Water Management Area (WMA). Historic and new weather stations (rainfall and temperature-logging sensors) are also indicated.

seldom produce large quantities of water, they are important both for local supplies and in supplying baseflow to rivers. Aerially limited quaternary age (alluvium cover) primary aquifers also exist in the northern sector of the catchment. Recharge is generally episodic, thus only occurring during intense rainfall events or during periods of prolonged rainfall. Based on historic data, groundwater is generally brackish. Salinity increases in the direction of groundwater flow and also with decreasing recharge.

Soils are generally poorly developed, shallow on hard or weathering rock, brownish sandy loams with lime generally present in most of the landscape. Some red and yellow soils with low to medium base status are also present. Soil depth usually varies between 0.5 m and 1 m, and the soil-water holding capacity is predominantly between 20 mm and 40 mm, but it can be up to 80 mm in the upper and lower reaches of the Sandspruit catchment. Soil drainage is somewhat impeded by the low hydraulic conductivity of the semi-weathered Malmesbury shale throughout the Sandspruit catchment and it is particularly poor in the lower reaches. Low to moderate swelling clays are also present, consisting mainly of kaolinite and mica (Rycroft and Amer, 1995). Soil salinity is moderate. The interflow component at the interface of soil cover and semi-weathered Malmesbury shale is prominent. Annual average runoff was estimated to be ~30 mm/a (De Clercq *et al.*, 2010).

The vegetation biome is fynbos. Much of the natural vegetation, i.e. *Swartland renosterveld* and *Hawekwas fynbos*, has been replaced by agricultural lands. Land use in the Sandspruit catchment is dominated by cultivated land and pastures. The catchment falls within the 'bread basket' of the Western Cape and thus agriculture is dominated by wheat cultivation. The growing of grapes, lupins and canola is also common. Farmers in the area generally follow a three-year planting rotation, i.e. cultivation only occurs every third year. Lands are left fallow between planting seasons and used for grazing. Soil erosion is minimised through the use of man-made anti-erosion contours, which are evident throughout the catchment.

The primary objectives of the monitoring programme in the Sandspruit catchment were to quantify the water balance and the refinement of the conceptual model for this catchment. The main users of the water-balance and conceptual model of the Sandspruit catchment were envisaged to be government departments (in particular DWA and the Department of Agriculture), local authorities (municipalities and CMAs), private entities (environmental impact assessment and consulting practitioners, farming communities, etc.), the general public as well as the scientific community. Applications of the water-balance and conceptual model were envisaged to be in water-resource assessment and planning (including state of the environment, water and chemical mass balance), but also in change detection (identifying short- and long-term trends). Additional applications were envisaged to be in hydrological modelling with distributed parameter models as well as in the development of water and pollution management strategies.

A full report on baseline data collected at Sandspruit can be found in Jovanovic *et al.* (2011). Collection of baseline data served the purpose of assessing what data were available and what monitoring gaps existed for a comprehensive description of the Sandspruit catchment. In particular, the following monitoring gaps were identified:

- *Atmospheric monitoring*: No weather station or known rain gauge was found within the Sandspruit watershed. The closest weather stations were at Moorreesburg and

De Hoek (South African Weather Services), Langgewens (Department of Agriculture and South African Weather Services) and Goedertrou (research weather station, De Clercq *et al.*, 2010) (Figure 5.2). It was therefore deemed necessary to install weather stations within the Sandspruit catchment. Rainfall and air temperature were recommended as the minimum weather measurements required to quantify the water balance. These variables are key inputs into hydrological models in the calculation of evapotranspiration and recharge/discharge.

- *Surface water monitoring*: Long records of water flow and chemistry data were available from DWA station No. G1H043, at the outlet of the Sandspruit catchment. However, no monitoring of sub-catchments was taking place.
- *Groundwater monitoring*: The information from the National Groundwater Archives (DWA) was used for the purpose of this study. However, it was deemed necessary to drill more boreholes for a comprehensive monitoring programme in the Sandspruit catchment for the following reasons:
  - borehole data (groundwater levels and chemistry) were erratic and sometimes only on a once-off basis;
  - borehole log data (both geological and geophysical) were not found and thus the vertical extent of aquifers and their hydraulic properties were not defined;
  - brackish groundwater was measured in close proximity to freshwater;
  - existing data had inconsistencies (e.g. the same groundwater data were measured at different times and locations).
- *Soil and vadose-zone monitoring*: No continuous monitoring of soil and vadose-zone parameters was taking place. Due to the geological nature of the catchment with soil and weathered material overlying Malmesbury shale characterised by low permeability, a strong interflow component was expected. It was therefore suggested to monitor this flow component through the installation of piezometers at the interface between the soil cover and the Malmesbury shale, where temporary perched water tables may occur especially during the rainy winter season. It was also suggested to collect sediment material during borehole drilling for laboratory analyses of water content, electrical conductivity (EC) and Cl concentrations. Electrical conductivity was measured to identify the origin and accumulation of salts in the vadose zone of this inherently saline area (De Clercq *et al.*, 2010), whilst Cl concentrations serve the purpose of estimating long-term evapotranspiration with the chloride mass balance method (Eriksson and Khunakasem, 1969) Water content of sediment samples was measured to convert EC and Cl concentrations in 1:5 solid:solution extracts into actual solution concentrations. Moreover, it was suggested to carry out resistivity measurements to indicate subsoil layering and water-flow pathways.
- A more detailed description of *land use* was also required, including the effects of man-made anti-erosion contours on water fluxes.

Three broad sections of the Sandspruit catchment were identified based on the geological characteristics:

1 Sandstone/Malmesbury shale geology in the upper reaches;
2 Undulated Malmesbury shale in the middle reaches; and
3 Malmesbury shale overlain by alluvial sandy soils in the lower reaches.

Each of these sections was investigated in detail and used as a basis for a more comprehensive monitoring programme. The geological characteristics are strongly linked to the type of soils present in the catchment, which in turn are associated with land uses (type of farming) in the area. Thus, in the design of the integrated catchment monitoring programme, the consequential effects of geology, climate, soil and land use were considered. In addition, the principle of integrated monitoring was adopted that includes collection of comprehensive data sets (weather monitoring, hydrometry, vadose-zone profiling, geophysical and isotope studies; Table 5.1). The purpose of data collection and the actions taken in the Sandspruit catchment are also shown in Table 5.1.

5.3.2 Expanded monitoring programme

In this section, examples of added value and benefits obtained from the expanded monitoring programme are presented for weather data, groundwater data, vadose-zone profiling, and geophysical data. Two years of data (2009–2010) were collected in the expanded monitoring programme aimed at filling the identified monitoring gaps and fulfilling requirements.

5.3.2.1 Methodologies

Atmospheric variables are drivers of the hydrological cycle. In particular, rainfall amounts and intensity affect hydrological processes like infiltration, runoff, drainage

Table 5.1 Data collection, purpose of data collection and actions taken to fill monitoring gaps in the Sandspruit catchment.

Data collection	Purpose of data collection	Action taken in the Sandspruit catchment to fill monitoring gaps
Weather	Rainfall and evaporation represent the driving force of water fluxes in the catchment	Establishment of three rain and temperature stations within the catchment
Hydrometry	Surface-water flow and quality measurements	Feasibility investigation on monitoring surface-water flow in sub-catchments (Jovanovic et al., 2011)
	Groundwater flow and quality measurements	Drilling of 24 boreholes along three cross-sectional transects
Vadose-zone profiling	Analyses of modes of water flow and contaminant transport in the sub-surface	Water content, EC and Cl analyses on disturbed sediment samples collected during borehole drilling
Geophysical study	Analyses of geological layering and water occurrences in subsoil	Resistivity measurements along three cross-sectional transects
Isotope studies	Analysis of flow paths and hydrograph separation	Sampling and analyses of ^{18}O and ^{2}H in groundwater and surface water (Bugan et al., 2012)

and the components of the hydrograph. Monitoring of weather variables is essential for estimating evapotranspiration and, indirectly, groundwater recharge. The four main atmospheric variables affecting evapotranspiration are air temperature, solar radiation, wind speed and relative humidity. In order to estimate reference evapotranspiration with the Penman-Monteith equation (Allen *et al.*, 1998), the minimum required weather data are daily maximum and minimum temperatures (Annandale *et al.*, 2002). Weather variables may vary greatly depending on the location within a catchment and particular topography. Three weather stations were installed within the Sandspruit catchment, i.e. in the upper reaches, middle reaches and in the lower reaches in the vicinity of the confluence with the Berg River. The new stations included temperature sensors installed in a Gill screen and automatic rain gauges. Data were logged at hourly intervals.

Three main transects were identified for the installation of boreholes across the slopes of the Sandspruit catchment. The three main transects were located in the upper, middle and lower reaches of the Sandspruit catchment (Figure 5.3). Each transect was established at a cross-section of the catchment, starting from the upper elevation down to the river bed. Transect 1 (Zwavelberg, Figure 5.3) in the upper reach, including 10 boreholes, was located on the lower slopes of the Kasteelberg Mountain. The geology in the area is characterised by TMG sandstone (towards the mountain), and alluvium/colluvium overlying Malmesbury shale on the opposite slope. Transect 2 (Oranjeskraal, Figure 5.3), including three boreholes, represents the middle reach of the Sandspruit, a Malmesbury shale-dominated environment. The location of transect 3 is also dominated by Malmesbury shale (Uitvlug farm, Figure 5.3, including five boreholes), but the alluvial cover is somewhat different in texture (generally sandier) compared to transect 2. Upon completion of the first three transects an additional transect was drilled (Malansdam farm, Figure 5.3, including four boreholes) representing the upper northern reach of the Sandspruit.

Drilling samples of 2 kg to 3 kg of representative sediment material were collected from layers (depths) that displayed characteristic features and different layering. The samples were sealed in sampling bags and used to measure soil-water content by drying sub-samples in an oven at a temperature of 105°C for at least 2 d. The sub-samples were subsequently used to prepare 1:5 solid:solution extracts. The sub-samples were first weighed, distilled water was added, the mixture shaken for about 45 min, left overnight and centrifuged for about 10 min. The resulting solution was used to measure EC and Cl concentration. Subsequently EC and Cl profiles were drawn. The upper reaches of the catchment coincide with the recharge area from the Kasteelberg sandstone formation and low levels of salinity were recorded there. In the middle and lower reaches of the catchment, high salinity levels were measured due to the presence of natural salts of meteoric origin trapped in sediments. It was deemed that this natural salinity would interfere somewhat with the interpretation of Cl profiles at these locations. Chloride profiles were therefore not described in the middle and lower reaches of the catchment. They were determined only in the upper reach (Zwavelberg, Figure 5.3).

Groundwater samples were collected for groundwater chemistry analyses on 6–8 September 2010. Samples were collected after purging three volumes of groundwater from boreholes, by filling sampling bottles to the top and sealing them to prevent contamination. The groundwater samples were handled according to the guidelines included in the ICM mind map and the samples were analysed for EC, pH and major cations and anions according to standard laboratory procedures. The depth to

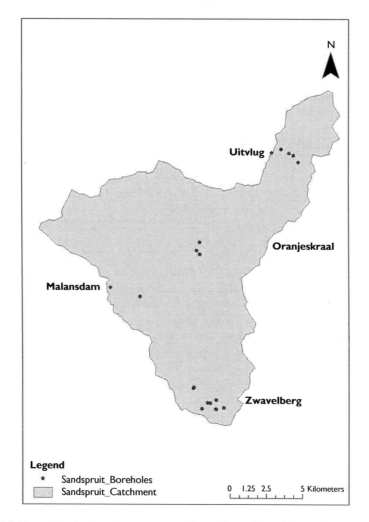

Figure 5.3 Map of the Sandspruit catchment with positions of boreholes and names of farms.

groundwater was also measured using a manual groundwater level meter (dip meter) during field visits.

Resistivity measurements were carried out at all borehole hill-slope transects (Figure 5.3) with a Lund imaging system. The resistivity tomography method was used to provide a pseudo-section of change in electrical properties in the subsurface along a specified line or transect. The bulk resistivity of different geological strata varies mostly because of changes in salinity of the pore fluid or changes in porosity of the host rock (Archie's empirical formula from Telford *et al.*, 1990). In our case, the aim of resistivity measurements was to detect changes in geological strata down to 15 m depth and along lines (transects) about 120 m in length. The data were collected using a standard protocol with the Wenner array. The apparent resistivity data acquired in the field were inverted using the RES2DINV software (Loke, 2001) to provide a true-depth resistivity section. The depth of the inverted section is often over-estimated in very conductive conditions. Depth-sounding data is then abstracted for a single representative lateral position on every profile to be able to adjust the depth to more realistic levels. A total of 12 resistivity transects were measured. Resistivity measurements took place between February and March 2009 (end of dry season).

5.3.2.2   Results

Measured rainfall tended to decline within the Sandspruit catchment from the upper reaches towards the lower reaches (from 494 mm/a to 321 mm/a in 2009) (Figure 5.4). Daily air temperatures exhibited an increasing trend from the upper reaches towards the lower reaches (Figure 5.5).

Figure 5.6 presents the gravimetric water content profile, chloride profile (Cl concentration solid:solution 1:5 ratio extracts) and the geological log of a borehole located on Zwavelberg farm in the upper reaches of the Sandspruit catchment (borehole No. ZB003). Peaks of Cl concentrations can be observed in the topsoil due to evaporation-driven accumulation of salts close to the surface, as well as at depths approaching saturation conditions (Figure 5.6).

The interpolated potentiometric surface across the Sandspruit catchment derived using historic data from the National Groundwater Archive was updated using data from the installed boreholes (Figure 5.7). The groundwater ECs measured at the newly installed boreholes were used to update the interpolated groundwater EC map, which was then compared to the historic interpolated EC map (Figure 5.8). The new groundwater EC map showed a more extensive area of high salinity in the downstream reaches and to the east of the catchment, compared to the historic map. Caution should, however, be exercised in the interpretation of interpolated point data of groundwater levels and EC. The piper diagram of the groundwater chemistry is shown in Figure 5.9. The following can be deduced from the chemical analyses:

- the quality of groundwater from the Table Mountain Group sandstone is good with EC ranging between 35 mS/m and 280 mS/m (Zwavelberg farm boreholes);
- groundwater salinity increases downstream. EC in groundwater ranged from 33 mS/m in the vicinity of the sandstone formation (Zwavelberg farm) to 2 200 mS/m at the Uitvlug farm underlain by the Malmesbury Group shales;
- a salinity hotspot was identified in the lower reaches of the catchment with groundwater EC ranging from 290 mS/m to 2 200 mS/m (Uitvlug farm);
- Na and Cl are the dominant ions present in groundwater at all boreholes. Magnesium, calcium and sulphate are occasionally elevated in the upper reaches of the catchment (Zwavelberg farm);
- very low concentrations of K were measured in all boreholes;
- the Sandspruit River is highly saline at DWA station No. G1H043 and NaCl is the dominant salt.

On the large scale, changes in lithology could be broadly delineated along resistivity profiles. Most of the area shows a higher general resistivity in deeper layers, which is due to the presence of fractured/weathered shale. For example, the high resistivity

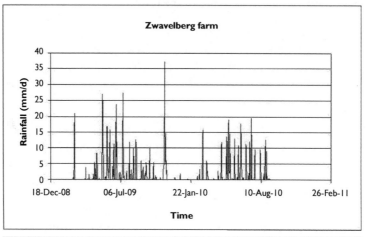

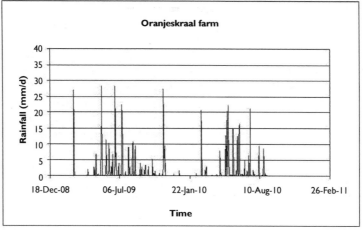

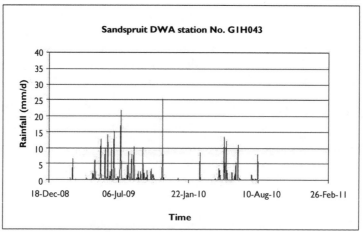

Figure 5.4 Comparison of daily rainfall at Zwavelberg (top), Oranjeskraal (middle) and Sandspruit DWA station No. G1H043 (bottom graph).

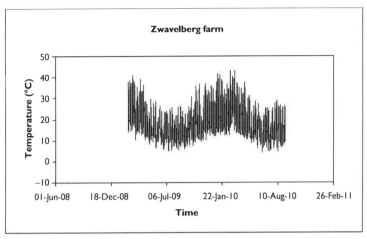

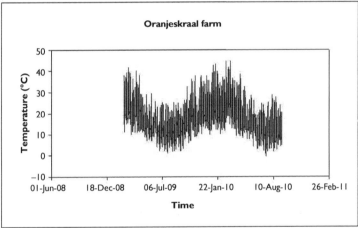

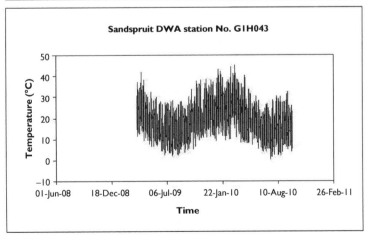

Figure 5.5 Comparison of daily air temperatures at Zwavelberg (top), Oranjeskraal (middle) and Sandspruit DWA station No. G1H043 (bottom graph).

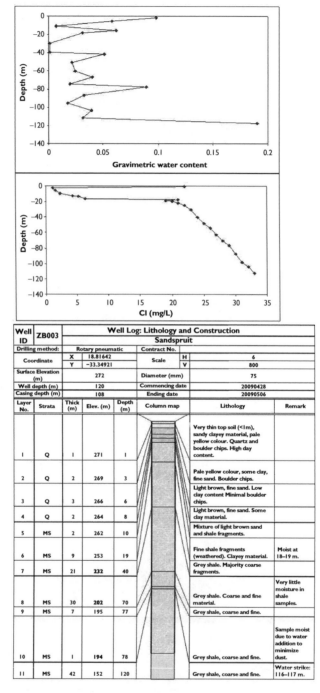

Figure 5.6 Water content profile (top, units in g water/g sediment), chloride profile (Cl concentration of solid: solution 1:5 extracts) (middle) and log description of borehole No. ZB003 at Zwavelberg farm (bottom, using Borehole Logging v. 1.0 Excel-based software).

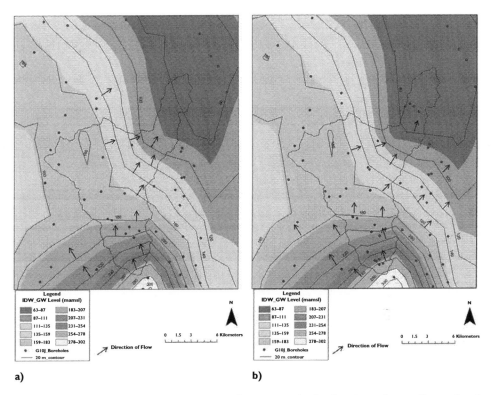

Figure 5.7 The groundwater potentiometric surface across the Sandspruit catchment (interpolated with inverse distance weighting) and the direction of groundwater flow drawn using (a) National Groundwater Archive data only (1924–2003); and (b) National Groundwater Archive and data from new boreholes (1924–2003 and 2009–2010).

deeper in the profile (Figure 5.10) is an indication of the presence of shale, which was verified during borehole logging and sampling (Figure 5.6). The high resistivity is due to the clayey texture of the material and very dry conditions, as measurements were taken at the end of the dry summer season.

5.3.3 Improved water balance and conceptual model

Data gathered during this investigation allowed for the annual water balance to be quantified and a conceptual flow model to be refined for the Sandspruit catchment (Bugan *et al.*, 2012). The catchment receives 473 mm/a precipitation on average. Higher rainfall (494 mm/a at the foot of Kasteelberg) was recorded in the upper reaches of the catchment where groundwater recharge mainly occurs through the sandstone fractured rock system, compared to the lower reaches (321 mm/a at DWA station No. G1H043). Streamflow at DWA gauge No. G1H043 was measured to be ~30 mm/a. Evapotranspiration makes up the remainder of the water balance (443 mm/a), assuming there are no other losses of water, e.g. regional groundwater losses directly through discharge into the Berg River. Soil water and groundwater storage are negligible

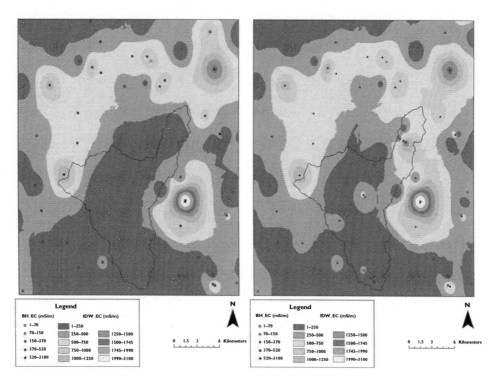

Figure 5.8 Electrical conductivity (EC) of groundwater map of the Sandspruit catchment (interpolated with inverse distance weighting) drawn using National Groundwater Archive data only (1965–2008, left graph); and National Groundwater Archive data and data from new boreholes (1965–2008 and 2009–2010, right graph).

components of the water balance in the long run. Seasonal fluctuations of the groundwater potentiometric surface suggested that evaporation impacts the groundwater table and that a seasonal groundwater recharge-discharge mechanism exists. The stream is seasonal and it is fed mainly through subsurface flow (interflow) during the winter rainy season. As groundwater recharge and discharge is less than streamflow (30 mm/a), the historic values of groundwater recharge of 69 mm/a to 71 mm/a estimated by Vegter (1995) for quaternary catchment G10J appear to be overestimated (assuming no other groundwater losses occur). The discrepancy between the current estimate of groundwater recharge and the values estimated by Vegter (1995) can be attributed to the data set and methodology used. Vegter (1995) indicated that groundwater recharge expressed in mm/a represents a long-term average and that recharge may occur occasionally every few years rather than annually, in particular in arid and semi-arid areas in the south-west of South Africa. Catchment processes like runoff, evapotranspiration and groundwater recharge depend on annual rainfall amounts and distribution. In this study, more recent rainfall time series were used compared to Vegter (1995) and changes in land uses and management may have caused less runoff and groundwater recharge. Vegter's (1995) groundwater recharge maps were based on point recharge results from other studies and they represent regional trends. In this

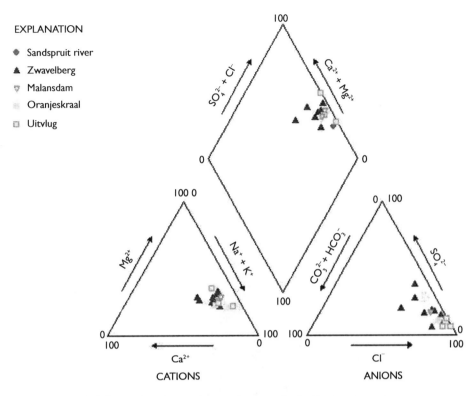

EXPLANATION

◈ Sandspruit river
▲ Zwavelberg
▽ Malansdam
▨ Oranjeskraal
▢ Uitvlug

Figure 5.9 Piper plot of groundwater chemistry in the Sandspruit catchment.

study, all data for the calculation of the water balance were collected within the Sandspruit catchment. In addition, a combination of experimental work and modelling, including hydrograph separation, was used to quantify the contributions to runoff from overland flow, interflow and baseflow (Bugan *et al.*, 2012).

A poor correlation between average annual streamflow and average rainfall ($R^2 < 0.4$) suggested that a variety of factors may influence streamflow, e.g. rainfall distribution, cropping systems and evapotranspiration, etc. Streamflow is therefore more dependent on the rainfall distribution in time than on annual rainfall.

The information from the expanded monitoring programme and the time series of data allowed us to better understand the system. The seasonal nature of the stream and the depth of the water table suggested that the regional groundwater contribution to streamflow is minimal, leaning towards negligible. Streamflow is driven by quickflow, which comprises overland flow and especially interflow from the alluvium cover. Temporary seasonal perched water tables occur at the interface of the alluvium cover and Malmesbury shale with low permeability, as identified in borehole logs during drilling. Infiltration is facilitated by preferential pathways created by root channels (e.g. winter wheat) as well as the minimisation of overland flow rates by the dense grass cover. In addition, man-made anti-erosion contours that are common in the area represent micro-areas where overland flow of water is barraged and

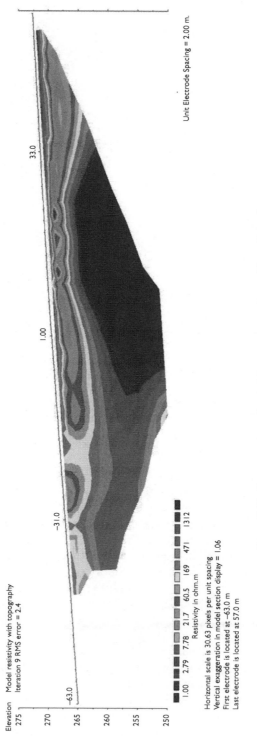

Elevation   Model resistivity with topography
275   Iteration 9 RMS error = 2.4

Resistivity in ohm.m

Horizontal scale is 30.63 pixels per unit spacing
Vertical exaggeration in model section display = 1.06
First electrode is located at –63.0 m
Last electrode is located at 57.0 m

Unit Electrode Spacing = 2.00 m.

Figure 5.10 Cross-section of resistivity measurements at Zwavelberg farm.

water infiltrates. The dominant contribution to the stream hydrograph is therefore interflow, originating from the recharge of temporary groundwater tables in winter (Bugan *et al.*, 2012).

5.4   CONCLUSIONS

This study confirmed the importance of establishing and maintaining sound water-monitoring programmes in catchments, with particular emphasis on the importance of monitoring the entire water cycle. This includes integration of all environmental compartments, namely groundwater, surface water, unsaturated zone and atmospheric measurements. The ICM mind map developed in this study includes guidelines and information on how to monitor the various compartments of the environment. An integrated monitoring approach is recommended (hydrometry, geophysics, isotope measurements, weather monitoring and vadose-zone profiling) where monitoring is done by specialists in the particular field. The main target users of the ICM mind map in South Africa are CMAs, but also government departments, private practitioners and water users as well as research institutions.

A number of recommendations on the roles and responsibilities in monitoring emanated from this study. National, provincial (catchment) and local governments may have tiered tasks. At the national level, water volumes and quality monitoring is the overall responsibility of the DWA as stipulated in Chapter 14 of the National Water Act of 1998. However, CMAs, water boards, municipalities or other local government organs, licensed users and WMAs have to monitor at provincial, regional catchment and local levels as all of them in one way or another are responsible for different aspects of water-quality protection and management. Therefore, results, guidelines and tools produced in this study are of interest at three levels of water governance, characterised by three different levels of data-network densities, namely local (municipalities and local institutions), regional (regional offices of DWA and/or CMAs) and national (DWA Head Office). The push toward development of CMAs in South Africa will change the use of data in such a way that these agencies also become primary users of the data collected in their catchments. The onus will therefore no longer be on DWA to manage and collect the entire country's data as DWA will provide a strategic context for water-quantity and -quality management. The scales of monitoring and interactions between governance levels need therefore to be regulated.

Specific recommendations refer to the modality of data exchange between institutions. Monitoring standards need to be adhered to in data collection and exchanges between custodians of databases. A specific case of interest is the one of rainfall monitoring and data exchange. The DWA runs a rainfall-monitoring system for specific objectives, in addition to the weather services network run by the South African Weather Services (SAWS). The SAWS data are available to the public upon request and they are easily obtainable by DWA. However, DWA data are not captured by SAWS mainly because of differences in monitoring standards. The ICM mind map supplies the web site of SAWS where guidelines on rainfall and weather data collection are available. A large amount of information, especially on rainfall and soils, exists and it needs to be collated and included in current databases. The sources of this information are universities, laboratories, private consultants, local government, schools, etc. It may be recommended that one of the government bodies (e.g. SAWS for weather, Provincial Departments of Agriculture for soils) should embark on sourcing and auditing the anecdotal information. Quality-control mechanisms for such data should be developed by experts. Such data could also be flagged for source, or their level of confidence could be added in the database.

The integrated monitoring guidelines were applied to a demo/pilot study site in the Sandspruit quaternary catchment. It was highlighted that there is a consequential cause-and-effect relationship between geology, climate, soil and land use. Baseline data were collected and a monitoring gap analysis was performed that led to the design of a more comprehensive monitoring programme. The benefits of the expanded monitoring programme were supported with data evidence and resulted in a better understanding of the natural system and in the development of an improved conceptual model and quantification of the water-balance fluxes.

The products and knowledge gained through this study fit into the broader programme of development of supporting tools to CMAs and other similar water-management boards. These can be part of an implementation programme where a toolbox could be made available to water managers. The ICM mind map can be easily expanded to update guidelines and include more guidelines as they are developed (e.g. guidelines on soil erosion and sediment monitoring, microbiological monitoring, etc.). Similarly, monitoring programmes should be seen as dynamic, as they may be updated, expanded and reduced as necessary. Ongoing refinement is possible through feedback loops between monitoring programmes and hydrological modelling.

ACKNOWLEDGEMENTS

The authors acknowledge the Water Research Commission for funding project No. K5/1849 from which this paper emanated, the National Research Foundation, DWA for funding the drilling of the boreholes, SA Rock Drill for drilling the boreholes and the farmers in the Sandspruit catchment for making their land available and for supplying valuable insight into the environmental conditions.

REFERENCES

Allen, R.G., Pereira, L.S., Raes, D. & Smith, M. (1998) Crop evapotranspiration: Guidelines for computing crop water requirements. Rome, Italy, United Nations Food and Agriculture Organization. Irrigation and Drainage Paper 56.

Annandale, J.G., Jovanovic, N.Z., Benade, N. & Allen, R.G. (2002) User-friendly software for estimation and missing data error analysis of the FAO 56-standardized Penman-Monteith daily reference crop evaporation. *Irrigation Science* 21, 57–67.

Bugan, R.D.H., Jovanovic, N. & De Clercq, W.P. (2012) The water balance of a seasonal stream in the semi-arid Western Cape (South Africa). *Water SA* 38, 201–212.

Canter, L.W., Knox, R.C. & Fairchild, D.M. (1987) *Groundwater Quality Protection.* Chelsea, MI, Lewis Publishers. p. 562.

Chapman, D. (1992) *Water Quality Assessments: A Guide to the Use of Biota, Sediments and Water in Environmental Monitoring.* London, UK, Chapman & Hall. p. 585.

Dallas, H.F. & Day, J.A. (2004) *The effect of water quality variables on aquatic ecosystems: A review.* Pretoria, South Africa, Water Research Commission. WRC Report No. TT 224/04.

De Clercq, W.P., Jovanovic, N. & Fey, M.V. (2010) *Land use impacts on salinity in Berg River water*. Pretoria, South Africa, Water Research Commission. WRC Report No. 1503/1/10.

DWAF (1997) *Minimum Standards and Guidelines for Groundwater Resource Development for the Community Water Supply and Sanitation Programme*. Pretoria, South Africa, Department of Water Affairs and Forestry.

DWAF (2004a) *Integrated Water Resources Management. Guidelines for Groundwater Management in Water Management Areas, South Africa*. Pretoria, South Africa, Department of Water Affairs and Forestry, DANIDA Funding Agency, March 2004.

DWAF (2004b) *Standard Descriptors for Geosites*. Pretoria, South Africa, Department of Water Affairs and Forestry.

DWAF (2006) *A Guideline for the Assessment, Planning and Management of Groundwater Resources within Dolomitic Areas in South Africa*. Edition 1. Pretoria, South Africa, Department of Water Affairs and Forestry.

DWAF (2007) *Artificial Recharge Strategy, Version 1.3*. Pretoria, South Africa, Department of Water Affairs and Forestry.

DWAF-DANCED (2007a) *Implementation of Existing Groundwater Quality Management Strategies*. [Online] Available from: www.dwaf.gov.za [Accessed 16th July 2007].

DWAF-DANCED (2007b) *Groundwater Monitoring and Integrated Monitoring Networks*. [Online] Available from: www.dwaf.gov.za [Accessed 16th July 2007].

Eriksson, E. & Khunakasem, V. (1969) Chloride concentrations in groundwater, recharge rate and rate of deposition of chloride in the Israel coastal plain. *Journal of Hydrology* 7, 178–197.

Jovanovic, N., Israel, S., Petersen, C., Bugan, R.D.H., Tredoux, G., De Clercq, W.P., Vermeulen, T., Rose, R., Conrad, J. & Demlie, M. (2011) *Optimized monitoring of groundwater – surface water – atmospheric parameters for enhanced decision-making at a local scale*. Pretoria, South Africa, Water Research Commission. WRC Report No. K5/1846.

Loke, M.H. (2001) Constrained time lapse resistivity imaging inversion. In: *Symposium on the Application of Geophysics to Engineering and Environmental Problems, SAGEEP*, 4–7 March 2001, Denver, CO, United States. Environmental and Engineering Geophysical Society. p. 34.

Nacht, S.J. (1983) Groundwater monitoring system consideration. *Groundwater Monitoring Review*, 3, 33–39.

Oakley, K.L., Thomas, L.P. & Fancy, S.G. (2003) Guidelines for long-term monitoring protocols. *Wildlife Society Bulletin* 31, 1000–1003.

Rycroft, D.W. & Amer, M.H. (1995) Prospects for the drainage of clay soils. Rome, Italy, Food and Agriculture Organization of the United Nations. Irrigation and Drainage Paper No. 51. p. 134.

Strobl, R.O., Robillard, P.D., Shannon, R.D., Day, R.L. & McDonnell, A.J. (2006) A water quality monitoring network design methodology for the selection of critical sampling points: Part I. *Environmental Monitoring Assessment* 112, 137–158.

Telford, W.M., Geldart, L.P. & Sheriff, R.E. (1990) *Applied Geophysics*. Cambridge, UK, Cambridge University Press.

UNEP/WHO (1994) Water quality monitoring – A practical guide to the design and implementation of freshwater quality studies and monitoring programmes. In: Bartram, J. & Balance, R. (eds.) *Water quality monitoring*. ISBN 0 419 22320 7 (Hbk) 0 419 21730 4 (Pbk). Geneva, Switzerland, United Nations Environment Programme and the World Health Organization. 348 pp.

Usher, B., Pretorius, J.A., Dennis, I., Jovanovic, N., Clarke, S., Cave, L., Titus, R. & Xu, Y. (2004) *Identification and prioritization of groundwater contaminants and sources in South Africa's urban catchments*. Pretoria, South Africa, Water Research Commission. WRC Report No. 1326/1/04.

Vegter, J.R. (1995) *An explanation of a set of National groundwater maps*. Pretoria, South Africa, Water Research Commission. WRC Report No. 74/95.

Chapter 6

Socio-economic aspects of groundwater demand: Franschhoek case study

D. Pearce[1], Y. Xu[1], E. Makaudze[2] & L. Brendonck[3]
[1]*Groundwater Research Group, Department of Earth Science,*
University of the Western Cape, Cape Town, South Africa
[2]*Department of Economics, University of the Western Cape, Cape Town,*
South Africa
[3]*Faculty of Natural Sciences, Laboratory of Aquatic Biology and*
Evolutionary Biology, University of Leuven, Belgium

ABSTRACT

This paper outlines the demand-side approach to the valuation of groundwater focusing on a case study conducted in Franschhoek, South Africa. Significant drivers of the demand for groundwater are also expounded.

The study is confined to the domestic sector of the sample area and utilises two approaches to ascertain the Total Economic Value (TEV) of groundwater in the research area; analysis of municipal water-billing archives to determine the consumer surplus of direct use value (market value); and contingent valuation to determine the non-market value. A survey was conducted, via face-to-face interviews, in order to facilitate the contingent valuation assessment. Demographic and perceptual household data were collected as well.

Economic expansion and immigration of people into the study area are identified as the most significant drivers of growth in the demand for groundwater for domestic consumption. Indeterminate results were obtained in the determination of the direct use value since the market did not show any significant response to changes in the price of water, as the price of water was too low to effect any significant change in consumption levels. Income levels, respondent age, municipal satisfaction levels, and Living Standard Measure (LSM) were identified as significant drivers of demand for the non-market values of groundwater. The contingent valuation produced a willingness-to-pay estimate for the non-market values of groundwater which was R498 125.00 per month for the research area.

6.1 INTRODUCTION

Groundwater is an essential part of the South African water market, though in the past, the role of groundwater has largely been confined to rural districts where the relatively small-scale use of has rendered surface-water schemes too expensive and to the agricultural sector where self-sourced groundwater is considerably cheaper than piped water purchased from the municipality. Woodford and Rosewarne (2006) estimate South Africa's total extractable groundwater resources to be approximately

19 000 Mm³/a, yet they estimate that only 1 770 Mm³/a is extracted annually, with 64% of that being used for irrigation.

However, the socio-economic changes that are now taking place in South Africa are driving an increasing interest in available freshwater resources (Braune and Xu, 2008). South Africa has a two-tiered economy with a highly developed formal sector rivalling other developed nations and an undeveloped informal sector characterised by significant unemployment and poverty (CIA, 2009). South Africa's major cities, with their modern infrastructure have thus become the epicentres of economic development whilst the rural regions remain relatively underdeveloped.

The contrast between the formal and informal sectors is driving a migration of rural populations that are increasingly headed towards urbanised centres in search of employment opportunities (Braune and Xu, 2010). The effect of this localised economic growth and the resulting population migrations is putting additional pressure on already strained water-supply systems, which consist almost completely of surface-water supplies in the form of dams. There is an urgent need to develop new supplies of freshwater, but with options for dam building all but exhausted, and desalinated water still being prohibitively expensive, groundwater is increasingly being recognised as one possible solution (TMG Aquifer Alliance, 2004; Xu et al., 2009).

In the Western Cape Province of South Africa groundwater resources represent a potentially significant source of available freshwater, but despite the pressure to develop new freshwater resources, investment in developing the potential of this resource has been slow. This can be attributed to the fact that there is still very little knowledge amongst policy- and decision-makers regarding the true value of groundwater (Lopi, 2009). This is due in part to the fact that these subterranean water sources are 'out of sight', which has resulted in the resource being relatively unnoticed and poorly understood (Braune and Xu, 2010). In general, there is a high level of ignorance amongst policy-makers as to the potential of this resource. As such, the role of groundwater has as yet not been adequately incorporated into institutional development frameworks, which in turn has translated into inadequate funding commitments.

In the case of groundwater, there is a specific need to understand the pressures that transition communities towards the active development of groundwater resources. The aim of this paper is thus to investigate the socio-economic value of groundwater using a demand-side approach and to identify the dynamics that underlie the demand for both the market and the non-market values of groundwater. The study assumes identical markets for the consumption of surface-water and groundwater resources in the context of municipal water use, founded on the argument that consumers are unable to distinguish the source at the final point of use, namely the tap. Only the peri-urban clusters are considered, and the study does not extend to the agricultural sector.

6.2 CONCEPTUAL FRAMEWORK

6.2.1 Conjunctive use

Limited water supplies are a potentially serious impediment to South Africa's economic expansion (DWAF, 2004) and as such there is recognition of the need to investigate and understand the nature of the growing demand for fresh water in South Africa's two-tiered economy. Although this study focuses primarily on groundwater, a degree of consideration must be paid to water use as a whole, which includes surface-water sources, as both contribute to the available supply. This simultaneous use of both surface-water and groundwater sources, without prior planning, is termed spontaneous conjunctive use (Foster and Steenbergen, 2011).

6.2.2 Legal framework

Legal obligation is helping to build the pressure to develop new water supplies. In South Africa the National Water Act of 1998 (DWAF, 1998a) provides the legislative framework under which water resources are managed and under this item of legislation priority is given to the provisioning of free basic water (25 l/person · d), a concept termed the human Reserve, as well as provisioning of water for the environment, a concept termed the ecological Reserve. Another important mandate within the act is the need to address the matter of equitable development and sustainable development (DWAF, 1998b), a matter which should only be addressed on a thorough understanding of consumer demand (Whittington et al., 1998).

6.2.3 Market vs. non-market values

In an economic sense there are two levels of value attributable to environmental factors; they are market values and non-market values. Market values are the values of environmental goods and services that are revealed through exchanges on the market whilst non-market values are the untraded values associated with an environmental resource (Barbier, 2005 and NRC, 1997).

In the case of groundwater an example of an associated market value may include the value of drinking water or water used for irrigation. In both of these examples the associated value is easily identifiable through market exchanges. In contrast, an example of a non-market value associated with an environmental factor is 'environmental quality' (Field and Field, 2002).

Economists often use the analogy of an iceberg to describe the relationship between market and non-market values. The submerged section of the ice represents the depth of consumer preferences and tastes and acts as the determinant and support for the visible tip of the iceberg represents the market values. The underestimation of these 'submerged' non-market market values is the source of much of the frustration surrounding the value of water resources, due to the fact that they are often far greater than anticipated and are regarded very dearly by those who experience them. This can be attributed to significant intrinsic, cultural, historical or religious factors (UNECA, 2006). The boundary between market and non-market goods is not stationary but rather, in our modern day economy, it is incredibly dynamic as more goods or services are either added to or removed from the actively traded market (Moss et al., 2003, NRC, 1997).

6.2.4 Combining valuation techniques

There is a reasonably well established body of literature in the field of environmental economics that makes use of a combination of complimentary techniques to investigate a range of environmental values associated with a water resource.

A recent example of this includes a study by Martinez-Paz and Perni (2010) who use a combination of CVM and a production-function approach to assess the benefits of different levels of groundwater protection. A study conducted in South Africa by Kanyoka *et al.* (2008) used a combination of two stated preference methods, specifically CVM and Choice Modelling (CM) to assess willingness to pay and consumer preferences for water services in rural areas. In a valuation of the economic consequences of different flow scenarios conducted by DWAF (2006), use was made of a 'Water Impact Model' which was itself a combination of methods designed to achieve an effective outcome.

Abdalla (1994) expresses the need for reduced complexity and for an integrated approach to such valuations. CVM has proved useful in this regard as it can be readily applied for a range of use and non-use values. Field and Field (2002) state that "virtually anything that can be made comprehensible to respondents can be studied with this technique" and in the informal African market it has proved to be an indispensible means to include impoverished communities in the decision making process (UNECA, 2006).

Direct approaches involving observable behaviour are always the preference, but these observable market values only ever account for a portion of the 'total economic value' spectrum.

6.3   DESIGN AND METHODOLOGY

6.3.1   Research area

Franschhoek is a small peri-urban community located in the South West of South Africa (illustrated in Figure 6.1) in the Western Cape Province. It is one of six wine regions that comprise the Cape Winelands District, the largest wine-growing region in South Africa. Franschhoek consists of several clusters of residential development nestled amongst farmland. The greater area, surrounded on three sides by mountainous terrain is known as the Berg River valley.

The area is located in the upper reaches of the Berg River catchment area, illustrated in Figure 6.2. Specifically, Franschhoek is located in quaternary catchment G10A.

The area was chosen and is suitable for this study for a number of specific reasons. Despite being located near several dams including the Skuifraam Dam, Theewaterskloof Dam and Wemmershoek Dam, Franschhoek is facing imminent water shortages. The water contained within these dams is proposed for the City of Cape Town (CoCT), which is facing its own looming water shortages. Franschhoek obtains approximately 66% of its water from Wemmershoek Dam (Stellenbosch Municipality, 2011a). This water was purchased from the CoCT under contract in 2001, but this contract has since expired and its renewal is not a certainty.

In addition, there is a significant contrast between the income levels of the wealthy and the poor and the areas playing host to the various income levels are geographically separate from one another, such that the lower income, lower middle income, upper middle income and upper income areas are distinct.

The two main clusters of development are Franschhoek Town and Groendal, with the latter located less than 3 km west of the former. Groendal consists of several sub-sections, namely Langerug, Mooiwater and La Motte, but for the purposes of this study they are recognised as a single collective entity. These two clusters are the focus of this study and comprise the extent of the research area.

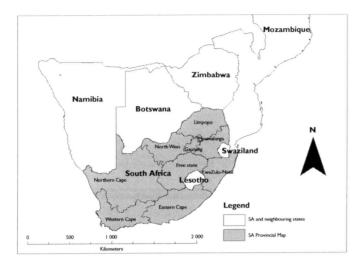

Figure 6.1 Map of South Africa indicating location of bordering countries and the research area.

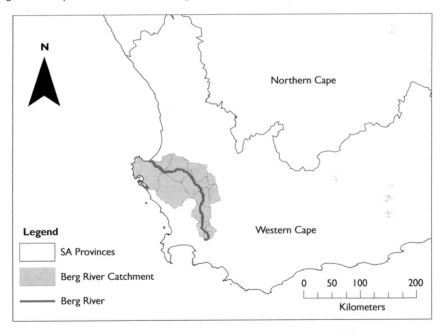

Figure 6.2 Map of the Berg River catchment (DWA, 2007).

Franschhoek Town and Groendal for the most part represent opposite ends of the economic spectrum. Franschhoek Town is comprised almost entirely of upper income and upper middle income households. Groendal residents cover the entire range of income levels; however, lower income and lower middle income households make up the majority of households in this area. A significant portion of the Groendal community is made up of informal dwellings.

6.3.2 Research design

The research design may be described as having three main components with each component purposed to collect a specific kind of data.

The first component is comprised of qualitative data collected by means of in-depth interviews conducted with industry experts. The purpose of the interviews was to obtain key pieces of information about the research area that would guide the complementary quantitative components. The areas of expertise explored in the interviews included water supply and reticulation, water-supply pricing structures, municipal procedures for archiving data, local environmental initiatives, water-supply history and recent economic developments.

The second component of the research design is comprised of longitudinal time series data documenting domestic water usage and billing within the research area. These time-series data were obtained from the local municipal archives and provided data relating to the quantity of water consumed per household per month and the associated cost, to the household, for the water that had been consumed. These data were used to identify patterns and trends in the total water consumption, water consumption per household and the price of water.

The third component is a survey purposed to provide cross-sectional data about the human population of the study area. The survey was used to conduct a contingent valuation of the non-market values of groundwater whilst also being used to collect data relating to the sample population including demographics, income, consumer preferences, etc.

The product of these three components allows identification of the major factors that underlie both the market and non-market socio-economic values of groundwater.

6.3.3 Residential survey design

The survey was administered via face-to-face interviews with residents in the research area. A randomly stratified sampling plan was utilised, with the geographic location being used as the primary variable for stratification. The research area is highly segregated with the geographic location of a particular residence being highly correlated with the income level and race of any given household.

The survey may be broadly defined as having two primary components. The first component consists of general survey questions whilst the second component relates to the contingent valuation of the non-market values of groundwater. The survey was conducted over a period of 6 d, three weekends to be specific. Weekends were chosen due to the increased likelihood of respondents being at home. Throughout the duration and upon the conclusion of the household survey the questionnaires were screened according to a pre-determined set of criteria to ensure the validity of the samples collected. A sample of 277 households was drawn from the population.

6.3.4 Contingent valuation design

The recommendations of Haab and McConnel (2003) served as the framework for the CV component of the residential survey. A contextual outline was designed with the input of a team of hydrogeologists and aquatic ecosystem specialists who were familiar with the research area. The contextual outline designed by the team followed the guidelines set forth in a report by the NRC (2004) as well as the recommendations of Turner *et al.* (2008) with regards to the ecosystem services approach and was purposed to describe the ecosystem services consumed by residents of the research area. It may be described as being comprised of the following elements:

- general information about groundwater and groundwater-dependent ecosystems in the region;
- descriptions of the benefits that groundwater and groundwater-dependent ecosystems generate for the residents of the research area;
- descriptions of the local activities that are currently having a negative effect on the local groundwater and groundwater-dependent ecosystems;
- the specific negative effect that the aforementioned activities are having upon the local groundwater resources and groundwater-dependent ecosystems.

In line with this contingent valuation process a hypothetical scenario was proposed to offset the degradation of local groundwater resources and to promote the ongoing viability of these resources. The proposed hypothetical scenario was centred on a state-run 'Groundwater Management Programme' (GMP) that would provide specific services to the community. The GMP would be organised and managed by the local municipality

The programme was designed around specific activities. The specific activities were to monitor the quality and availability of the resource; to identify sources of contamination; manage clean-up actions where necessary; organize legal actions where necessary; promote increased community awareness; ensure the sustainable use of groundwater; and promote its ongoing viability as a water resource. In essence the GMP was thus designed to be a proxy for all of a full range of non-market values of groundwater, including indirect use value, existence value and bequest value. This was done such that willingness to pay for the GMP would constitute willingness to pay for the non-market values of groundwater.

The hypothetical Groundwater Management Programme was described as being a 5-year programme with a total projected cost of R7 million rand and would be funded by means of a levy that residents of the research area would be charged. The levy would be included in the monthly water bill, and would be sustained for the 5-year period of the GMP. Upon conclusion of the programme, its renewal would be put up for consideration.

In line with other municipal pricing schemes, the GMP would use a stepped tariff charging a slightly different levy depending on the income level of the household. The contingent valuation utilised 2 different bidding amounts to simulate this stepped tariff. The initial bid proposed for upper income and upper middle income households was R80, whilst an initial bid of R40 was proposed for those residents in the middle income, lower middle income and low income categories. These amounts, given the size of the communities, along with a small state contribution would be sufficient to provide the R7 million needed to fund the 5-year GMP.

A critical component of the contingent valuation design is the method of asking the questions to elicit the amount that the respondent is willing to pay for the hypothetical scenario. Table 6.1 below, illustrates the sequence of questions used to elicit these values.

Table 6.1 Questionnaire structure to elicit 'willingness to pay'.

42. Do you understand the situation thus far? (If the respondent says 'No', explain the hypothetical scenario again.)	Yes	No
43. Would you vote 'Yes' or 'No' for the groundwater management program? (If the respondent says 'Yes', go to Q.44, otherwise skip to Q.45)	Yes	No
44. Would you be willing to pay an amount of R_____. (If 'Yes' iterate upwards until response is 'No') (If 'No' iterate downwards until response is 'Yes')	Yes	No

First Bid [   ]   Final Bid [   ]

45. If you voted 'No', for the program, what are your reasons for doing so?

This is government's responsibility, I should not have to pay extra for this.	
I don't trust government to run a program like this.	
I can't afford to pay for this program.	
Other	

Other _____
(Specify)

The initial elicitation of willingness of WTP was recorded as the 'first bid'. Dichotomous choice format is used. If the respondent said 'Yes' then the bid was iterated upwards until an amount was reached that the respondent was unwilling to pay. If the respondent said 'No' then the amount was iterated downwards until an amount was reached that the respondent was willing to pay.

6.3.5   Determining the parameters and estimating WTP

Binary logistic analysis is used to analyse the relationship between survey data and the yes/no responses of the respondents in order to identify significant factors that might influence the probability of respondents answering either 'Yes' or 'No' for a given bid.

The model is constructed as the concatenation of a composite income term (*CompositeIncome*) and a covariate matrix z. The composite income term and covariates presented in Table 6.3 and Table 6.5 were constructed according to the instructions of Haab and McConnel (2003) where:

$$CompositeIncome = ln\left(\frac{y_j - t}{t}\right)$$

y_j = Household Income for the jth respondent
t = bid term

The composite income term is thus the log of the ratio of the residual income (household income minus the bid price) to the bid price proposed in the hypothetical scenario.

A bound logit model is used to estimate the value of WTP. The parameters for this model are obtained from the output of the binary logistic analysis.

The bound logit is defined as follows by Haab and McConnel (2003):

$$WTP = \frac{\bar{y}}{1 + \exp(-\bar{z}\gamma)}$$

ZY can be calculated using the coefficients in the covariate matrix which are estimates of y/σ, the coefficient of the composite income term which is an estimate of $1/\sigma$ and the mean values of the covariates which are the mean values for z such that:

$$\bar{z}\gamma = \gamma_0 + z_1\gamma_1 + z_2\gamma_2 \cdots + z_n\gamma_n$$

6.4   RESULTS AND DISCUSSION

6.4.1   Economic development in the area

The Franschhoek area's wine-making heritage and beautiful natural setting, combined with its proximity to the city of Cape Town and the rapid growth of the South African tourism economy have seen it expand from a quiet country community into a booming tourism destination in less than 20 years.

The high demand for property in the area, particularly in Franschhoek Town, has seen a dramatic increase in property prices and an ongoing series of housing development projects (Stellenbosch Municipality, 2011a). The number of active water-consuming stands in Franschhoek Town has increased from 756 units in March 2005 to 1 066 units February 2010; an increase of 41% over a 5-year/60-month period. This demand is being driven by local interest as well as foreign investors who seek to settle permanently or semi-permanently in the area. The semi-permanent foreign residents usually occupy their properties during the warmer summer months and return home during the winter. They have become known as 'the swallows', in reference to their annual migration. The tourist influx in the summer months gives the area an exaggerated seasonal demand for potable water with the seasonal peak being 300% greater than its winter low.

In addition, there has been a significant increase in informal settlement in the form of unskilled labour that has migrated into the area in search of employment in the hospitality industry. Municipal estimates of the size of the Langerug Community (the informal housing component of Groendal) were approximate 800 informal dwellings in November 2008, 1 200 informal dwellings in March 2010 and 1 700 informal dwellings in April 2011, an increase of 112% over a 41-month period.

6.4.2   Water sources

The Franschhoek area obtains its raw water from two primary sources, namely Wemmershoek Dam and the Mont Rochelle Nature Reserve (Stellenbosch Municipality,

2011a). Water is abstracted from several points within the Mont Rochelle Nature Reserve. Water is abstracted from the Du Toits River at two points, a pump station and a weir. The Du Toits River is primarily fed by interflow from subterranean water sources and as such is categorised as groundwater for the purposes of this study. Additional water is supplied from perennial spring-water-fed streams.

Table 6.2 below provides a breakdown of water usage in the research area, including the agricultural sector. Groundwater is actively abstracted on the local farms and the quality is such that it is used for both irrigation as well as a drinking-water source.

Water from the Wemmershoek Dam was being provided under contract from the City of Cape Town. The contract expired several years back and is yet to be renewed. This supply continues, but without any formal agreement.

According to 2010 figures 0.51×10^6 m^3 were supplied from the Mont Rochelle sources whilst 0.99×10^6 m^3 were supplied from Wemmershoek Dam. Thus approximately 34% of municipal water consumed in the Franschhoek Valley originates from groundwater resources.

6.4.3   Water usage pattern and trends

On the whole, water consumption for domestic use has been increasing over the past five years. This has been consistent despite the annual increases in the price of water. In Table 6.3 (below) water consumption figures are given for Franschhoek Town over five years (five periods of one year each) from March 2005 through to February 2010. Over this time total per annum water usage has increased from 313 787 m^3 in Period 1 to 444 384 m^3 in Period 5, corresponding to an average annual growth rate of 9.09%.

Consumption per stand has shown some growth over the five-year period from March 2005 through to February 2010, increasing from 32.80 m^3 per stand in Period 1 up to 36.84 m^3 per stand in Period 5. This increase in per stand consumption accounts for approximately 37% of the growth in consumption. The increase in the number of stands from 792 stands in Period 1 up to 1 002 stands in Period 5 accounts for the remaining 63% of the growth in consumption.

Table 6.2 Water-use breakdown for Franschhoek (Stellenbosch Municipality, 2011a; Stellenbosch Municipality, 2011b).

Use	mcm	%
Farm GW Drinking	0.13	2.31
GW Irrigation	1.00	17.10
SW Irrigation	3.19	54.85
Municipal GW (Mont Rochelle)	0.51	8.75
Municipal SW (Wemmershoek Dam)	0.99	16.99
Total	5.82	100

where:
  SW = Surface Water
  GW = Groundwater
  mcm = million cubic meters.

Table 6.3 Water-consumption data for Franschhoek Town.

Franschhoek Town					
Period	Mar'05–Feb'06	Mar'06–Feb'07	Mar'07–Feb'08	Mar'08–Feb'09	Mar'09–Feb'10
Total Consumption	313787	349186	396780	411813	444384
Total Cost (Rand)	1239808.48	1500286.76	1747675.53	2234857.44	2715502.58
Rand/m³	3.75	4.19	4.32	5.31	5.59
m³/stand	32.80	32.96	35.99	35.21	36.84
# of active stands	792	883	920	975	1002
Eco Growth %	Baseline	6.275	5.225	1.675	−1.175
Inflation %	Baseline	4.98	7.69	11.33	8.99
Inflation Adjusted					
Real Total Cost (Rand)	1239808.48	1518793.82	1728736.92	2018923.67	2224336.38
Real Rand/m³	3.75	4.24	4.27	4.80	4.87

The price of water in Franschhoek Town increased from R3.75/m³ in Period 1 up to R5.59/m³ in Period 5, corresponding to an average annual growth rate of 10.5%. The real value of money, however, is not constant. If the figures are adjusted to account for economic growth and inflation, with Period 1 as the baseline, then it is clear that the average cost per cubic meter of water increased from R3.75 in Period 1 up to R4.87 in Period 5. This is a significant increase and equates to an average annual growth rate of 6.75%.

Basic economic theory states that there is an inverse relationship between the price of a good and the quantity demanded of that good (Hardisty and Ozdemiroglu, 2008) However, as can be seen from the data above, despite the relatively sharp increase in the price of water, there was no reduction in per stand consumption of water.

In the case of Groendal the situation is a little more complicated. Groendal is divided into three main areas, namely formal housing, government subsidised formal housing and informal housing. Municipal water-billing data are archived only for water that is purchased from the municipality. Thus data are available to represent water consumption in the formal housing sector but none for the informal sector where water is provided free of charge. Of the 1 400 plus dwellings in the formal housing sector of Groendal approximately 700 are government subsidised houses.

Water-consumption data in the formal housing sector of the Groendal community are given in Table 6.4.

Total water consumption for the Groendal cluster grows steadily from 239 253 m³ in Period 1 to 306 804 m³ in Period 5 at an average rate of 6.41% per annum. Growth in the size of the cluster over this period was minimal as the number of active stands increased by 50 units over the 5-year period. Per stand consumption increased at an average rate of 5.51% per annum from 14.16 m³ per stand in Period 1 up to 17.55 m³ in Period 5. Thus 87.46% of the increase in water consumption may be attributed

Table 6.4 Water-consumption data for Groendal.

Groendal					
Period	*Mar'05–Feb'06*	*Mar'06–Feb'07*	*Mar'07–Feb'08*	*Mar'08–Feb'09*	*Mar'09–Feb'10*
Total Consumption	239253	281970	273744	297999	306804
Total Cost (Rand)	792209.67	7977971.99	984030.47	1228994.59	1374021.11
Rand/m³	3.29	3.50	3.61	4.17	4.52
m³/stand	14.16	16.48	15.88	17.14	17.55
# of active stands	1407	1426	1437	1449	1457
Eco Growth %	4.95	6.275	5.225	1.675	−1.175
Inflation %	3.58	4.98	7.69	11.33	8.99
Inflation Adjusted					
Real Total Cost (Rand)	792209.67	990035.94	973367.07	1110248.12	1125495.21
Real Rand/m³	3.29	3.55	3.57	3.76	3.71

to increases in per stand consumption with only 12.54% being attributable to the increase in the number of stands.

The average price of water increased from R3.31/m³ in Period 1 up to R4.91/m³ in Period 5, corresponding to an average annual increase of 10.36%. Accounting for economic growth and inflation over the 5-year period produces a real price growth rate of 5.04% per annum up to the price of R4.03/m³ in Period 5.

Municipal estimates for water consumption in the informal housing sector of the Groendal Community place it at approximately 9 000 m³/month as of April 2011. With approximately 1 700 dwellings in the informal settlement, and an average of 4.16 persons per informal dwelling (survey estimate), that amount correlates to approximately 42 l/person · d. This amount fits the consumption approximations given by the Department of Environmental Affairs of less than 50 l/person·d in informal housing settlements (DEA, 1999).

6.4.4 Non-market values

The samples collected were analysed to determine whether there was significant response bias. Of the 191 interviews that were collected from the less affluent segment of the community, 44 responded that they would not vote for the hypothetical scenario. Of the 86 interviews that were collected from the more affluent segment, 30 stated that they would not vote for the hypothetical scenario. The number of respondents who responded positively to the hypothetical scenario totalled 203, with 147 and 56 positive responses from the less affluent and more affluent market segments, respectively. The cumulative 'No' response rate was 27%, which represents a weak but tolerable level of response bias within the sample.

The nature of the response bias was somewhat different for each of the segments. In the less affluent segment, 40.9% of those who voted 'No' for the hypothetical scenario stated that they would not be able to afford it. These responses were most common amongst those living in informal dwellings. In the more affluent segment, the stated reasons for voting 'No' did not tend towards any specific issue. Binary logistic regression was used to analyse the data in order to identify significant covariates that might influence the probability that a respondent will respond either positively or negatively to a certain bid price. The respondent's response to the bid price (Yes/No) was used as the dependent variable. As different bidding prices were used for each of the segments, the data were analysed separately.

The binary logistic regression model for the more affluent market segment is presented below in Table 6.5.

CompositeIncome is also the only variable in the model that is significant at the 5% level. The dummy variables *Satisfaction_Dummy* and *Edu_Dummy* with Wald scores of 2.000 and 2.592 are slightly significant. The coding for the 4 dummy variables *Satisfaction_Dummy*, *Edu_Dummy*, *Race_Dummy* and *RentOwn_Dummy* is illustrated below in Table 6.6.

The positive beta of 1.093 for *CompositeIncome* indicates that the probability of a positive response to a certain bid price is directly correlated to the income level of the household. A negative beta for the dummy variable *Satisfaction_Dummy* indicates that people are less like to respond positively to a given bid if they are dissatisfied with municipal water services. A somewhat surprising finding is the negative beta for *Edu_Dummy* which indicates that respondents with degree-level qualifications or higher are actually less likely to respond positively to a given bid.

The model for the less affluent segment was a fair bit more stable, most likely due to the larger sample of 147 interviews. The binary logistic model for the less affluent market segment is presented in Table 6.7 below.

Table 6.5 Outputs of the binary logistic model for the 'more affluent' market segment.

Variables in the equation

	B	S.E.	Wald	df	Sig.	Exp(B)
CompositeIncome	1.093	0.484	5.095	1	0.024	2.983
Satisfaction_Dummy	−2.447	1.730	2.000	1	0.157	0.087
Edu_Dummy	−1.612	1.001	2.592	1	0.107	0.199
Race_Dummy	0.864	0.830	1.084	1	0.298	2.373
RentOwn_Dummy	−0.872	0.776	1.265	1	0.261	0.418
Constant	−1.892	2.162	0.765	1	0.382	0.151

Table 6.6 Dummy variable coding for the 'more affluent' market segment.

Variable	0	1
Satisfaction_Dummy	Satisfied with municipal water services	Unsatisfied with municipal water services
Edu_Dummy	Up to diploma	Degree and higher
Race_Dummy	Non-white	White
RentOwn_Dummy	Owns the dwelling/property	Renting the dwelling/property

Table 6.7 Outputs of the binary logistic model for the 'less affluent' market segment.

Variables in the equation

		B	S.E.	Wald	df	Sig.	Exp(B)
Step 1	CompositeIncome	0.817	0.347	5.533	1	0.019	2.263
	LOGage	−2.568	0.909	7.975	1	0.005	0.077
	Senior_Dummy	−1.517	0.599	6.420	1	0.011	0.219
	Satisfaction_Dummy	−1.163	0.577	4.069	1	0.044	0.313
	Dwelling_Dummy	1.494	0.612	5.951	1	0.015	4.454
	OwnRent_Dummy	−0.753	0.745	1.021	1	0.312	0.471
	Marital_Dummy	0.564	0.486	1.344	1	0.246	1.757
	Constant	6.783	3.441	3.887	1	0.049	883.148

Table 6.8 Dummy variable coding for the 'less affluent' market segment.

Variable	0	1
Senior_Dummy	No one older than 55 yrs	Has at least 1 person older than 55 yrs
Satisfaction_Dummy	Satisfied with municipal water services	Unsatisfied with municipal water services
Dwelling_Dummy	Informal Housing	Formal Housing
OwnRent_Dummy	Renting the property/dwelling	Owns the property/dwelling
Marital_Dummy	Not married	Married

The term *CompositeIncome* is constructed according to the same criteria set out in the model for the more affluent segment. *LOGage* May be quite simply defined as the log of the age of the respondent. The two continuous variable *CompositeIncome* and *LOGage* are significant at the 5% and the 1% levels respectively. Three of the five dummy variables, namely *Senior_Dummy, Satisfaction_Dummy* and *Dwelling_Dummy* are significant at the 5% level. The coding for all five of the dummy variables is presented in Table 6.8 shown below.

The positive beta for *CompositeIncome* indicates a positive correlation between the income level of a given household and the likelihood that they will respond positively to a given bid price. The negative beta for *LOGage* tells of a negative correlation between respondent age and the likelihood that he/she might respond positively to a given bid price. The dummy variable *Senior_Dummy* supports this finding as its negative beta indicates that households with at least one member over the age of 55 years are less likely to respond positively to a given bid price. *Satisfaction_Dummy*'s negative beta indicates that respondents who are not satisfied with municipal services are less likely to respond positively to a given bid price whilst the positive beta for *Dwelling_Dummy* indicates that those living in informal housing are less likely to respond positively.

6.4.5   Estimating WTP

The various parameter estimates for the less affluent segment are presented in Table 6.9.

The parameter estimates for the less affluent segment are presented below in Table 6.10.

The estimate of median WTP for the less affluent segment, based on the figures presented in Table 6.11 is R119.77 per household where $\sigma = 1.22399$ and $\bar{z}\gamma = -3.9976$

Table 6.9 Parameter estimates for the 'more affluent' market segment model.

More affluent model		(n = 56)		
			Parameter	Mean
Household Income			y	27424.3
Bid Term			t	80
	Parameter	Value		
CompositeIncome	$1/\sigma$	1.093		
Constant	γ_0/σ	−1.892		
Satisfaction_Dummy	γ_1/σ	−2.447	z_1	0.84
Edu_Dummy	γ_2/σ	−1.612	z_2	0.47
Race_Dummy	γ_3/σ	0.864	z_3	0.63
RentOwn_Dummy	γ_4/σ	−0.872	z_4	0.34

Table 6.10 Parameter estimates for the 'less affluent' market segment model.

Less affluent segment		(n = 147)		
			Parameter	Mean
Household Income			y	6643.53
Bid Term			t	40
	Parameter	Value		
CompositeIncome	$1/\sigma$	0.817		
Constant	γ_0/σ	6.783		
LOGage	γ_1/σ	−2.568	z_1	3.6791
Senior_Dummy	γ_2/σ	−1.517	z_2	0.24
Satisfaction_Dummy	γ_3/σ	−1.163	z_3	0.81
Dwelling_Dummu	γ_4/σ	1.494	z_4	0.7
OwnRent_Dummy	γ_5/σ	−0.753	z_5	0.81
Marital_Dummy	γ_6/σ	0.564	z_6	0.48

Table 6.11 Summary of 'willingness to pay' estimates.

WTP estimates	
Less Affluent Segment $\sigma = 1.22399$ $\bar{z}\gamma = -3.9976$	R119.77 per household
More Affluent Segment $\sigma = 0.9149$ $\bar{z}\gamma = -3.53547$	R456.84 per household

The estimate of median WTP for the more affluent segment, based on the figures presented in Table 6.11 is R456.84 per household where $\sigma = 0.9149$ and $\overline{z}\gamma = -3.53547$

These results are summarised in Table 6.11 below.

6.5   DISCUSSION AND CONCLUSIONS

In the South African context groundwater is still a very misunderstood and underutilised resource. It is quite clear that as demand for new supplies of potable water increase, so the demand for groundwater will increase accordingly, but what this paper demonstrates is that the socio-economic factors that drive this demand for water/groundwater vary at different levels within the socio-economic spectrum.

In the informal housing sector of Groendal, the total demand for water is driven purely by the size of its population, since access to water is provided free of charge. The informal sector has seen significant growth as rural populations migrate to urban areas in search of better employment prospects and improved access to services. Urbanisation, quite evidently, is the most significant factor here.

The setting is quite different for the formal housing sector of Groendal, where increases in the demand for water have been significant despite only a limited increase in the number of stands. Here, more than 87% if the growth in the demand for water is attributed to the increase in per stand consumption. The increased prevalence of backyard dwellings is arguably one of the drivers behind this increase, since this translates to a higher number of individuals living on a specific property and greater income for the owners of the property. It is also quite plausible that household sizes have increased over the past few years as households absorb extra tenants or 'family members' in search of employment in the area.

The highest end of the economic spectrum is defined by large development, with Franschhoek Town being the site for several sizeable high-income housing projects. The increase in the number of stands accounts for approximately 63% of the increase in the demand for water. The increase in water consumption per stand may most likely be attributed to two factors. Firstly, the vast majority of the new houses being built in the area have relatively large plots with sizeable gardens. These would most likely have a higher than average consumption of water which would drive up the average in the area. Secondly, as new housing developments are completed, it takes some time for all of the units to be occupied by their new owners, leading to a slight delay in consumption increases. Unoccupied units would in this case not necessarily lower the consumption average, since most of the new houses are inside gated communities with their own in-house gardening services. Thus even when houses are unoccupied a significant amount of water would be used.

Indeterminate results were obtained in the calculation of the direct value of groundwater due to the fact that per household water consumption is irresponsive to increases in the price of water. In fact, the data clearly indicate that per stand consumption is increasing in all sectors. The increases in the price of water were thus too low to trigger any reduction in consumption. This prohibited any attempts at deriving a demand curve which might have been used to measure consumer surplus. Fraiture and Perry (2007) argue that this is not uncommon in water markets, where below a certain threshold price the market is irresponsive to any changes in the price of water. This phenomenon is used to support the argument that the price of water may too low.

It was, however, established that 34% of water consumed in the research area originates from groundwater resources which generated revenue of approximately R2.2 million in 2010. Estimations of the non-market value of groundwater were more successful. In the more affluent segment of the Franschhoek community, the average willingness to pay for the non-market values of groundwater was calculated at R456.84 per household, for the 66% of households that were willing to vote for the Groundwater Management Programme. In the less affluent segment of the Franschhoek community, the average willingness to pay for the non-market values of groundwater was calculated at R119.77 per household, for the 77% of households that were willing to vote for the Groundwater Management Programme. With the more affluent segment of 970 stands, and a less affluent segment of 2 230 stands, total willingness to pay for the Franschhoek area is estimated at R49 8125.00 per month.

ACKNOWLEDGEMENTS

The support of the Water Research Commission of South Africa (WRC), provided in the form of both funding and expertise for this project, is hereby greatly acknowledged and appreciated. The support of Professor Luc Brendonck of the University of Leuven, in Belgium, is hereby acknowledged. In addition, we would like to acknowledge the University of the Western Cape, South Africa, and the Flemish Interuniversity Council, Belgium, for their contributions towards conducting this study.

REFERENCES

Abdalla, C. (1994) Groundwater values from avoidance cost studies: Implications for policy and future research. *American Journal of Agricultural Economics* 76, 1062–1067.

Barbier, E. (2005) *Natural Resources and Environmental Economics*. 1st edition. New York, NY, Cambridge University Press.

Braune, E. & Xu, Y. (2008) Groundwater management issues in Southern Africa – An IWRM perspective. *Water SA* 34, 699–706.

Braune, E. & Xu, Y. (2010) The role of groundwater in sub-Saharan Africa. *Groundwater* 48, 229–238.

CIA (2009) *South Africa – CIA – The World Factbook*. Langley, VA, Central Intelligence Agency.

DEA (1999) *Human Settlements and the Environment*. Pretoria, South Africa, Department of Environmental Affairs.

DWAF (1998a) *National Water Act*. Pretoria, South Africa, Department of Water Affairs and Forestry.

DWAF (1998b) *Guide to the National Water Act*. Pretoria, South Africa, Department of Water Affairs and Forestry.

DWAF (2004) *National Water Resource Strategy*. Pretoria, South Africa, Department of Water Affairs and Forestry.

DWAF (2006) *Komati Catchment Ecological Water Requirements Study: Valuation of Socio-economic Consequences of Flow Scenarios.* Compiled by Tiou and Mallory. Pretoria, South Africa, Department of Water Affairs and Forestry.

DWAF (2007) *Berg River Baseline Monitoring Project: Groundwater Specialist Report.* Compiled by Parsons and Associates, Bellville, Cape Town, South Africa.

Field, B. & Field, M. (2002) *Environmental Economics: An Introduction.* 3rd edition. New York, NY, McGraw-Hill.

Foster, S. & Steenbergen, F. (2011) Conjunctive groundwater use: A 'lost opportunity' for water management in the developing world? *Hydrogeology Journal.* [Online] Available from: doi: 10.1007/s10040-011-0734-1. [Accessed September 2011].

Fraiture, C. & Perry, C. (2007) Why is agricultural water demand irresponsive at low price ranges? In: Molle, F. & Berkhoff, J. (eds.) *Irrigation water pricing: The gap between theory and practice.* Wallingford, UK, and Colombo, CABI Publishing and International Water Management Institute, p. 12.

Haab, T. & McConnell, K. (2003) *Valuing Environmental and Natural Resources: The Econometrics of Non-market Valuation.* Cheltenham, UK, Edward Elgar Publishers.

Hardisty, P. & Ozdemiroglu, E. (2008) *The Economics of Groundwater Remediation and Protection.* Boca Raton, FL, CRC Press.

Kanyoka, P., Farolfi, S. & Morardet, S. (2008) Households' preferences and willingness to pay for multiple use water services in rural areas of South Africa: An analysis based on choice modelling. *Water SA* 34, 715–723.

Lopi, B. (2009) Policymakers should understand the value of groundwater. Pambazuka News 437. [Online] http://www.pambazuka.org/en/category/advocacy/56910 [Accessed 11th June 2009].

Martinez-Paz, A. & Perni, J. (2010) Environmental cost of groundwater: A contingent valuation approach. *International Journal of Environmental Research* 5, 603–612.

Moss, J., Wolff, G., Gladden, G. & Guttieriez, E. (2003) Valuing water for better governance. Business and Industry CEO Panel for Water. [Online] http://www.pacinst.org/reports/valuing_water/valuing_water_paper.pdf [Accessed December 2009].

NRC (1997) Valuing ground water: Economic concepts and approaches. Washington, DC, National Academy Press. [Online] http://www.nap.edu/catalog/5498.html [Accessed December 2009].

NRC (2004) Valuing ecosystem services: Toward better environmental decision-making. Washington, DC, National Academy Press. [Online] http://www.nap.edu/catalog/11139.html [Accessed December 2009].

Stellenbosch Municipality (2011a) *Water Services Development Plan. Bellville, South Africa.* Compiled by WorleyParsons SA, Consultants, Cape Town, South Africa.

Stellenbosch Municipality (2011b) *Reconciliation Strategy for Franschhoek, La Motte and Groendal. Stellenbosch, South Africa.* Compiled by Umvoto Consultants, Cape Town, South Africa.

TMG Aquifer Alliance (2004) Background information document for the Table Mountain Group aquifer (TMG) feasibility study and pilot project. [Online] http://www.tmgaquifer.co.za/docs/BIDMarch04 [Accessed February 2010].

Turner, K., Georgiou, S. & Fischer, B. (2008) *Valuing Ecosystem Services: The Case of Multi-functional Wetlands.* London, UK, Earthscan.

UNECA (2006) *Water for Sustainable Socio-economic Development.* Addis Ababa, Ethiopia, United Nations Economic Commission for Africa.

Whittington, D., Davis, J. & McClelland, E. (1998) Implementing a demand driven approach to community water supply planning: A case of Luguzi, Uganda. *Water International* 23, 134–145.

Woodford, A. & Rosewarne, P. (2006) *How Much Groundwater Does South Africa Have?* Compiled by SRK Consulting, Cape Town, South Africa.

Xu, Y., Jia, H. & Lin, L. (2009) *Groundwater flow conceptualization and storage determination of the Table Mountain Group (TMG) aquifers.* Pretoria, South Africa, Water Research Commission. WRC Report No. 1419/1/09.

Chapter 7

Using numerical modelling to cope with uncertainty in the real world: Examples from various models in the Western Cape

Helen Seyler & E.R. Hay
Umvoto Africa (Pty) Ltd, Muizenberg, Cape Town, Western Cape, South Africa

ABSTRACT

Groundwater practitioners are often challenged with questions like: how sure are you of the sustainable yield? What will be the impact of future changing rainfall patterns? How sure are you that my borehole will not be impacted? Numerical modelling is a powerful tool to address these questions and deal with uncertainties both in the data, and in the conceptual understanding of the data. This is shown with several models from the Western Cape of South Africa.

Numerical modelling for the Water Availability Assessment Study of the Berg Water Management Area provided information to support pre-feasibility planning decisions, such as groundwater availability, and the potential impact of large-scale abstraction on surface-water systems and on ecologically sensitive environments.

A regional model in the Hermanus area, for the municipal supply well field, shows that the magnitude of recharge has a significant effect on the groundwater system, supporting the notion that gaining accuracy in recharge numbers is an important activity. Conversely, the testing of various conceptual possibilities for the recharge pathway shows that this has minimal impact on the water levels at the well field.

Relying on a combination of good conceptual analysis with sound modelling techniques, models can go a long way to provide quantitative answers to key management questions.

7.1 THE UNCERTAINTY ROADBLOCK

Groundwater practitioners are often challenged with questions like: how sure are you of the sustainable yield? What will be the impact of future changing rainfall patterns? How sure are you that my borehole will not be impacted? We are often faced with the perception that there is not enough certainty of the numbers or the groundwater system for a licence to be awarded, or for interested and affected parties to give their buy-in, or for the landowner to allow you to drill a municipal production hole on their land. The various sources of uncertainty are:

- in the conceptual model;
- in recharge numbers and patterns;
- in the hydraulic parameters;
- in the fracture networks;
- over the possible impact of abstraction, on other users, and on the environment.

In each of these situations and for each of these possible sources of uncertainty, numerical modelling is a powerful tool to address these questions and to deal with uncertainties both in the data and the key parameters, and in the conceptual understanding of the data.

Various modelling examples from the Western Cape of South Africa are presented below, following the order of a typical life cycle of a water resource project, from pre-feasibility stage to full-scale implementation, and well field management. At each stage in this cycle numerical models can be used to address several of the uncertainties listed above.

7.2 CASE STUDIES

Quantifying the unexploited available groundwater resources was one aim of the Water Availability Assessment Study of the Berg Water Management Area, conducted for the Department of Water Affairs (DWA), and one area of groundwater interest was the alluvial aquifer of the Breede River basin (Figure 7.1). A numerical model was used to quantify the volume of water available for abstraction, and to provide support over the uncertainty of the potential impact abstraction may have on the hydraulically linked surface water system.

A multi-layered 3D regional model in MODFLOW was set up. Scenario testing showed that when abstracting 80% recharge, the system stabilises to a new steady state

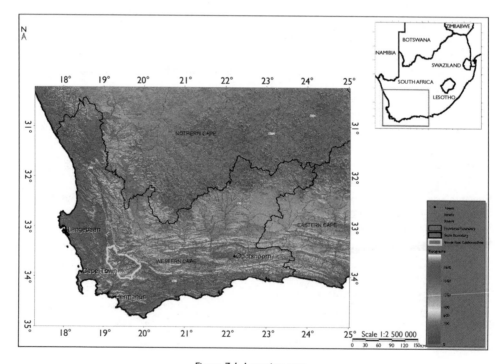

Figure 7.1 Location map.

water balance within a timescale of around 10 years (see Figure 7.2). As time passes under the new recharge regime, influx to the alluvial aquifer from the surrounding mountain springs increases ('springs' in Figure 7.2), and discharge from the aquifer to the Breede River decreases. The largest changes in these fluxes occur in the first 5 years, and by 10 years the system has largely re-stabilised; illustrated by the imbalance in Figure 7.2 coming close to zero (DWAF, 2008a).

Concern and uncertainty over potential environmental impact is often a 'show stopper' in the early stages of a project, or it can be a costly (in terms of time and money) hurdle to overcome. Numerical modelling adds important information to the debate, providing a quantitative response to questions over the degree of impact.

Figure 7.3 shows the modelled water level distribution, over long time scales, for a hypothetical well field 10 km from an ecologically sensitive environment (the Langebaan Lagoon, Western Cape, Figure 7.1). A cross-section of the water levels along the marked line of section shows that the well field would impact regional water tables by a few metres (see the difference between the 'natural system' and the 'well field' water levels at 10 km in Figure 7.4). The water levels at the lagoon (20 km) are, however, un-impacted (DWAF, 2008b).

In the preliminary stages of a water resource development project, uncertainty in the hydrogeological system is effectively handled with a regional model, constructed to test various conceptual models. The impact of various uncertainties on the system

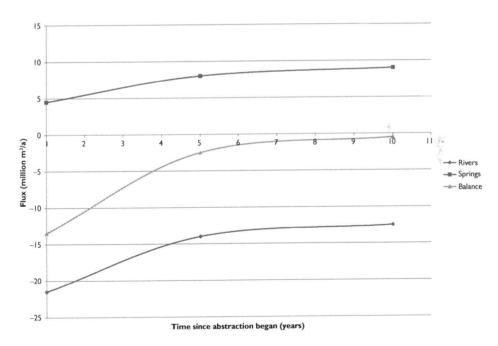

Figure 7.2 Modelled fluxes of the Breede River Alluvial Aquifer when subjected to 80% recharge, for time since large-scale abstraction commenced. Fluxes into the aquifer are positive, discharges are negative. A negative balance shows more water leaves the aquifer than enters in that year (DWAF, 2008a).

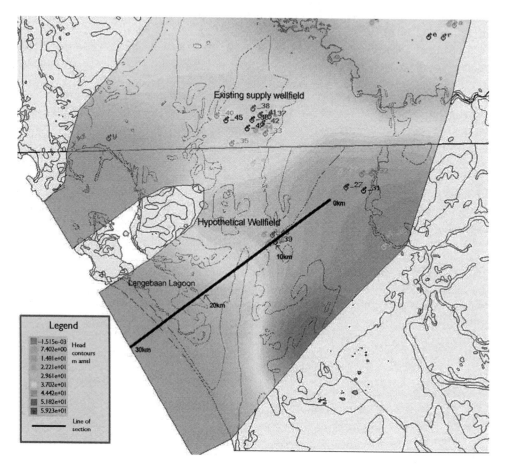

Figure 7.3 Modelled water level distribution for of the Langebaan Road and Elandsfontein Aquifer System, shown for a hypothetical well field scenario (DWAF, 2008b).

can be estimated, and thus the characterisation of these uncertainties (which can require costly data collection), can be prioritised.

Figure 7.5 shows the conceptual model for the Peninsula Aquifer at the Gateway well field, used for municipal supply, in Hermanus, Western Cape (Figure 7.1). This is translated into a multi-layered 3D regional model in MODFLOW, to test various conceptual models for recharge pathways. Some key results are:

1   As expected, recharge volume has a significant impact on water levels: hence accurate determination of recharge is a key requirement to implementation of the Gateway well field.
2   The recharge pathway (focused along the fault, or dispersed across the north-eastern aquifer boundary) has a negligible impact on the water levels, and hence detailed geological mapping to address the uncertainties in the deep Peninsula structure is less of a priority (Umvoto, 2009).

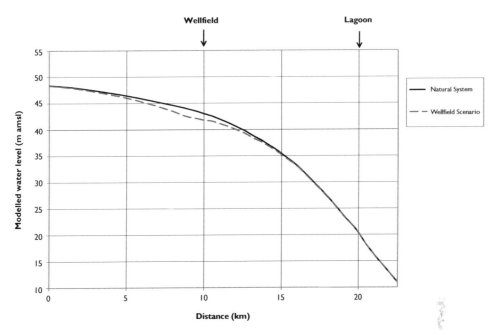

Figure 7.4 Modelled water levels with distance along line of section for of the Langebaan Road and Elandsfontein Aquifer System, comparing modelled water levels for the natural system and for a hypothetical well field scenario (DWAF, 2008b).

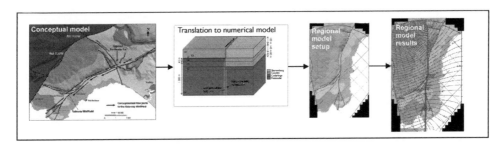

Figure 7.5 Graphical depiction of the process for development of the regional model for Gateway well field, Hermanus, from conceptual model, to numerical model design, model set-up and model results (Umvoto, 2009).

For complex aquifer systems where analytical solutions are not appropriate, questions such as how the well field will respond to an alternate pumping scenario, or to increased abstraction in drought periods can often only be accurately addressed with numerical modelling. The model is then a useful tool to support decision-making over operating rules.

Operation of the Gateway well field, Hermanus, is supported with a well field model that builds on the regional model. The model is constructed in the finite element modelling package FEFLOW (DHI-WASY GmbH, 2010). Table 7.1 below shows calibration statistics of the well field model, for the steady-state water levels. The difference between modelled and observed water levels ranges from 0.0 m to a maximum of 0.61 m.

Confidence in a model is built through testing it against data which have not as yet been used in the calibration process: 'validation and verification' (i.e., comparison without further calibration to these new conditions). Figure 7.6 shows how the Gateway well field model matches water levels for an operational pumping data set. Although there are some discrepancies, for example the difference in water levels over longer time recovery, the magnitude of the drawdown is well matched. This result has been effective in building confidence in the ability of the model to evaluate responses to conditions that are as yet not tested.

Table 7.1 Model calibration statistics for the Gateway well field (Umvoto, 2011).

Well	Water level (m amsl)		Error
	Observed	Modelled	
GWP01	33.82	33.95	0.13
GWP02	33.40	33.73	0.33
GWM05	39.05	38.94	−0.11
GWP06	34.05	34.66	0.61
GWE09	33.64	33.64	0.00
GWE10	33.31	33.39	0.08
	Absolute residual mean =		0.21
	RMS/quadratic mean =		0.09
	Sum of residuals =		1.03

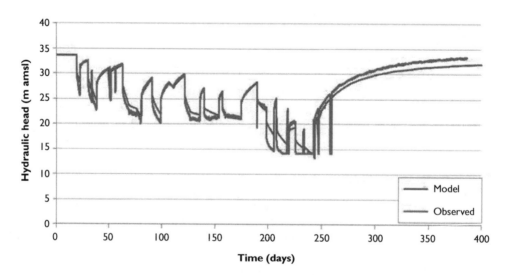

Figure 7.6 Comparison of modelled and observed water levels for a validation data set of irregular operational pumping, Gateway well field Model (Umvoto, 2011).

The model fluxes for the same operational pumping model run are shown in Figure 7.7. The 'imbalance', the sum of all fluxes, is negative during the first 200 d of pumping when recharge is low, indicating that more water is taken from the aquifer than enters, and water is taken from aquifer storage. During this time the natural discharge to the ocean gradually decreases. The appropriate question in terms of impact is whether this reduction in flux is environmentally acceptable. Before abstraction ceases at around day 250, the 'imbalance' is positive as the volume of recharge is greater than the abstraction, and aquifer storage is replenished.

Due to its proximity to the coast one of the key sources of uncertainty over the sustainability of long-term use at the Gateway well field, is the risk of saline water intrusion. The well field model was extended to a multi-density flow model, again using FEFLOW, to explore the various factors controlling saline intrusion, to test the impact of the geological structures on saline intrusion, and to test the salinity response to over-abstraction (Von Scherenberg and Seyler, 2012). Figure 7.8 shows the salt concentration, (above background), and water level for pumping since zero time. The pumping causes a minor rise in background concentrations, and the model scenario suggests that the salinity would rise from Class 0 (0 mS/m to 70 mS/m) to Class 1 (70 mS/m to 120 mS/m) over very long timelines. A drought situation is reflected at around 3 600 d, where twice the recharge volume is abstracted for 1 year. The concentration shows that a rise in salinity beyond acceptable levels does not happen, and the concentration returns to the background salinity levels once over-pumping ceases (Von Scherenberg and Seyler, 2012).

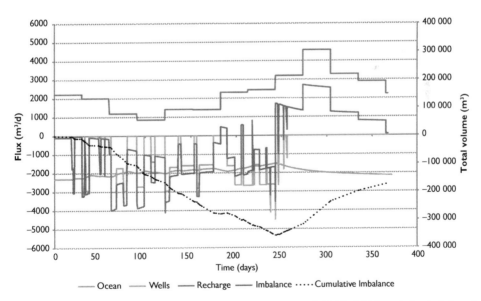

Figure 7.7 Model fluxes for the validation data set, Gateway well field Model. 'Cumulative imbalance' plots on the secondary axis, all other series on the primary axis. Positive fluxes are flows into the modelled aquifer, negative ones are outflows (Umvoto, 2011).

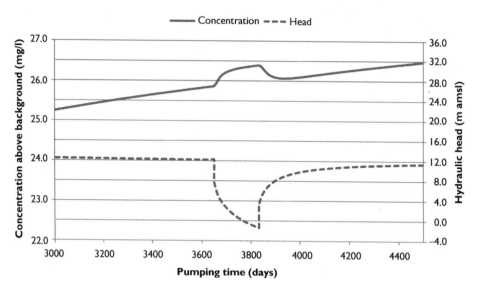

Figure 7.8 Modelled salt concentration and water level response over time to an over-pumping scenario (Von Scherenberg and Seyler, 2012).

7.3 SUMMARY

The case studies described above illustrate that modelling at the early stage of a project can effectively move a project from the uncertainty realm into action. Modelling during management phase takes you from acting retroactively based on monitoring data, into a practice of modelling to evaluate future, potential changes in the aquifer thus informing management decisions, and monitoring to see how closely the model was able to simulate the events, thus leading to the fine-tuning of a model before embarking on another cycle of evaluations.

The examples presented here of course do not address the possible uncertainty within the numerical modelling, on which there is a large amount of literature and research focusing on mathematical methods to quantify the uncertainty of the model result. However, few of these are able to consider the uncertainty in the hydrogeological structure. The examples given here show that through a combination of good conceptual analysis and sound modelling techniques, models can go a long way to provide quantitative answers to key management questions.

ACKNOWLEDGEMENTS

This study draws on various projects undertaken by Umvoto Africa Pty (Ltd). The valued support by the DWA Directorate: National Water Resource Planning during the Berg Water Availability Assessment Study project was key for the successful inclusion of numerical groundwater modelling into the project. The on-going support of the Overstrand Municipality is also acknowledged.

REFERENCES

Department of Water Affairs and Forestry, South Africa (2008a) The assessment of water availability in the Berg Catchment (WMA 19) by means of water resource related models: groundwater model report: Volume 8 – Breede River alluvial aquifer model. Prepared by Seyler, H., Hay, E.R., Hartnady, C.J.H. and Riemann, K., of Umvoto Africa (Pty) Ltd in association with Ninham Shand (Pty) Ltd on behalf of the Directorate, National Water Resource Planning. DWAF Report No. PWMA 19/000/00/0408.

Department of Water Affairs and Forestry, South Africa (2008b) The assessment of water availability in the Berg catchment (WMA 19) by means of water resource related models: groundwater model report: Volume 6 – Langebaan Road and Elandsfontein aquifer system model. Prepared by Seyler, H., Hay, E.R., Hartnady, C.J.H. and Groenewald, L., of Umvoto Africa (Pty) Ltd in association with Ninham Shand (Pty) Ltd on behalf of the Directorate. National Water Resource Planning. DWAF Report No. PWMA 19/000/00/0408.

DHI-WASY GmbH (2010) FEFLOW 5 Software. Available via FEFLOW. [Online] http://www.feflow.info/uploads/media/user_manual_01.pdf [Accessed 12th October 2010].

Umvoto Africa (2009) Water source development and management plan for the greater Hermanus area: Numerical groundwater modelling for the Hermanus Well field. Prepared by Seyler, H. and Dodman, A., of Umvoto Africa (Pty) Ltd on behalf of the Overstrand Municipality. June 2009.

Umvoto Africa (2011) Water source development and management plan for the greater Hermanus area: Numerical groundwater modelling for the Gateway Well field 2. Prepared by Seyler, H., Von Scherenberg, L., and Webb, M., of Umvoto Africa (Pty) Ltd on behalf of the Overstrand Municipality. February 2011.

Von Scherenberg, N.L. & Seyler, H.G.P. (2012) Assessing the impact of saline intrusion with density dependent flow modelling for the fractured Peninsula Aquifer in Hermanus, South Africa. *Water Science and Technology: Water Supply* 12, 387–397.

Chapter 8

Numerical modelling techniques for fractured aquifers and flooding of mines

K.T. Witthüser & C.M. König
Delta-H (Pty) Ltd, Pretoria, South Africa

ABSTRACT

This paper summarises statistical methods to generate stochastic fracture networks and modelling strategies for underground mines, both of which can be used for the quantitative risk assessment in the Water Systems Modelling Software SPRING. Fracture parameters gathered in the field need to be corrected for a number of sampling biases like geometric error of length bias before the statistical parameters are derived. Using statistical parameters, stochastically generated discrete fracture network models can be incorporated into finite-element models, where three-dimensional rock-matrix elements are linked by their nodes to two-dimensional fracture elements. A specific meshing technique allows an accurate representation of heterogeneous flow and transport processes in the fractures and matrix of secondary aquifers. The software models the hydraulic impacts of mine flooding on a regional scale by appropriate discretisation of mine voids and an accurate representation of layered fractured aquifer systems with numerous free groundwater surfaces. Relevant flooding processes within a mine occur on different spatial and temporal scales compared to the regional flow regime. Temporal changes of hydraulic parameters due to the progress of mining and the long-term groundwater depletion in the vicinity of the mine must be taken into account.

Keywords: numerical modelling, fractured rocks, stochastic fracture networks, mine flooding

8.1 INTRODUCTION

A quantitative risk assessment in fractured rocks often requires the consideration of discrete fractures (used here to mean any type of discontinuity in a rock mass, e.g. joints, fractures, faults or bedding planes) in numerical groundwater flow and transport models as the hydraulic dominant feature(s) and main pathway(s) for potential pollutants. Depending on the scale of the model, fracture network generation can be treated deterministically as discrete fracture sets or stochastically if the scale of observation and the number of fractures becomes larger. A stochastic approach requires that statistical fracture parameters have to be evaluated from field measurements. The evaluation of statistical parameters is performed by fitting statistical distribution models like log-normal or exponential distributions to empirical field data. Beyond the rather straightforward task of generating stochastic discrete fracture networks, their implementation in numerical models requires advanced mesh-generation techniques, as described below.

8.2 DETERMINATION OF INPUT PARAMETERS

8.2.1 Fracture mapping

Statistical fracture parameters can be determined using the scan-line technique, an objective sampling technique where all fractures intersecting a survey line attached to an outcrop are sampled. While the technique allows a very accurate determination of statistical fracture parameters (orientation, trace length, spacing and aperture), the associated time and cost implications should not be underestimated. The geometric error (Terzaghi, 1965) or the underrepresentation of fractures striking slightly oblique to a scan-line (the same is applicable to boreholes) can be minimised as far as possible by aligning the scan-lines preferably perpendicular to the dominant fracture sets and by sampling fracture data in various orientations with comparable scan-line lengths. Despite these efforts, the sampling bias still needs to be statistically corrected. Since the number of fractures intersected by a scan-line is directly proportional to the sinus of their intersection angles, the inverse was used as a weighting factor to compensate for the geometric error (Figure 8.1). An upper weighting threshold of 10 (Priest, 1993) should be applied to avoid potential overrepresentation of single fracture sets striking more or less parallel to a scan-line.

8.2.2 Orientation of fracture cluster

The fracture orientations are plotted as contoured polar diagrams in the lower hemisphere using the cosine-exponent-weighting-method, which ensures a minimal grid contouring bias (Adam, 1989). The clusters or sets of preferred orientations are determined by identifying their respective maximum densities (Figure 8.2) and choosing an appropriate selection angle, typically around 20°, for each cluster. A symmetrical Fisher (i.e. spatial normal) distribution is assumed for each cluster:

$$f(\theta, \alpha) = \frac{k}{4\pi \sinh(k)} \exp(k\cos\theta) \tag{8.1}$$

where:
θ is the angle between the mean direction (pole) and the observed value
α the dip, and
k the concentration parameter

The spherical variance ω is related to the concentration parameter k via:

$$\hat{\omega} = \arcsin\sqrt{2\frac{1-\frac{1}{m}}{\hat{k}}} \tag{8.2}$$

While the visual identification of fracture sets assumes a symmetrical Fisher distribution of fracture sets, in comparison to a mathematical cluster algorithm, it allows a geological and theoretical (measurement error) interpretation of observed variances.

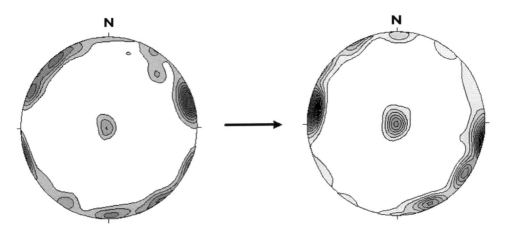

Figure 8.1 Statistical correction of the geometric error.

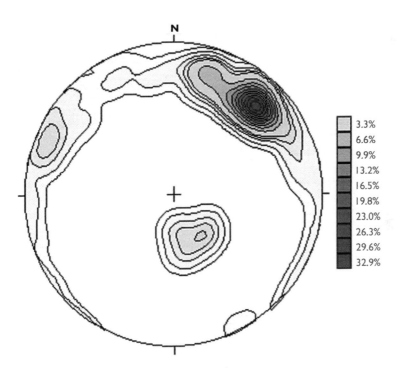

	3.3%
	6.6%
	9.9%
	13.2%
	16.5%
	19.8%
	23.0%
	26.3%
	29.6%
	32.9%

Figure 8.2 Example of a contoured polar diagram of two orthogonal fracture sets and sub-horizontal bedding planes in a sandstone outcrop.

Once the fracture sets are selected, their mean direction $\alpha N/\varphi N$ and spherical variance/apertures ω are calculated. While the absolute orientations of fracture sets in the folded sedimentary rocks vary and do not allow for a regional extrapolation, the relative orientation between the fracture sets and bedding planes typically remains constant and enables the generation of stochastic fracture networks within deterministically described bedding plane orientations.

8.2.3   Fracture trace length

Any fracture survey on a two-dimensional plane (i.e. outcrop) is limited to the determination of fracture trace-length distributions instead of real fracture sizes. Fracture sizes can be statistically derived from an observed trace-length distribution if a fracture geometry (e.g. circular or quadratic) is assumed. Before such a deduction can be made, the fracture trace-length distribution needs to be corrected for considerable sampling bias (Kulatilake *et al.*, 1993). The truncation bias, a result of omitting fractures below a minimal trace length, is negligible if a sufficiently short cut-off length (e.g. 0.05 m) is chosen. The censoring bias accounts for the fact that not all terminations of a fracture are necessarily visible in an outcrop. A prime example are multi-layer fractures in sedimentary rocks, which do not terminate against bedding planes and whose limits are therefore often partially (1-sided) or completely (2-sided) concealed in limited outcrop areas. Chilés and de Marsily (1993) provide a statistical correction based on the complete as well as 1- and 2-sided censored trace-length distributions. However, despite substantial data-gathering efforts only insufficient censored data (below 1% of all gathered data) were collected in three different fracture-mapping projects with thousands of mapped fractures to merit the calculation of censored trace-length distributions. The error was therefore considered negligible.

The most important sampling error, the size or length bias, arises from the fact that the probability of intersecting a fracture in an outcrop is directly proportional to its length. It results essentially in an overrepresentation of larger fractures, while neglecting smaller fracture sets which might contribute substantially to the overall connectivity of a fracture network. If an exponential distribution is assumed for the true (circular) fracture length, the theoretically observed fracture trace-length distribution can be derived. The observed fracture trace-length distribution $g(l)$ is directly proportional to the length l and the unbiased fracture length distribution $f(l)$ with the constant of proportionality c (Priest and Hudson, 1981):

$$g(l) = clf(l) \tag{8.3}$$

Using the definition of the statistical mean $E_f[l]$

$$g(l) = \frac{l}{E_f[l]} f(l) \tag{8.4}$$

and an assumed exponential probability density function (pdf) for $f(l)$

$$f(l) = \lambda_L \exp(-\lambda_L l) \tag{8.5}$$

the Erlang-2 distribution is derived for the trace-length distribution observed in an outcrop:

$$g(l) = \lambda_L^2 l \exp(-\lambda_L l) = \frac{l}{E_f[l]^2} \exp\left(-\frac{l}{E_f[l]}\right) \qquad (8.6)$$

By fitting the observed trace-length distribution with the Erlang-2 distribution (Figure 8.3), the parameter of the unbiased 'true' exponential distribution can be derived. Neglecting the size bias can result in a twofold overestimation of the average fracture trace length:

$$E_f[l] = \frac{1}{\lambda_L} = \frac{E_g[l]}{2} \qquad (8.7)$$

While the Erlang-2-PDF is very similar to frequently cited lognormal-PDFs for trace length, it enables a direct estimation of the unbiased distribution parameter and elimination of size bias.

8.2.4 Fracture spacing and density

The fracture intersections along a scan-line might follow a purely random Poisson process or be spatially correlated in e.g. so called war zone models (Black, 1994), with increased fracture densities next to major faults. If the fracture intersections are spatially independent (Poisson process), the observed fracture spacing along a scan-line should follow an exponential distribution (Chilés, 1988) and a random distribution

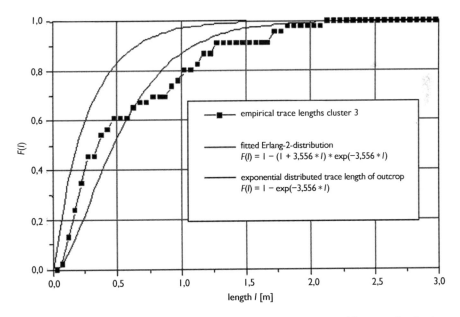

Figure 8.3 Example of a fracture trace length and derived exponential fracture distribution.

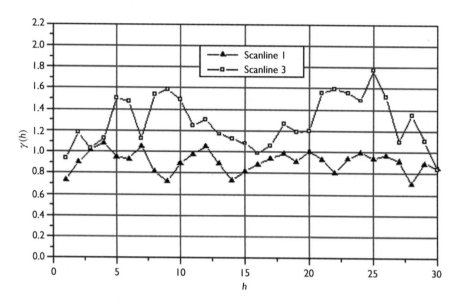

Figure 8.4 Example of fracture density variograms.

of fracture centres can be applied for the generation of the discrete fracture network. In addition, variograms of fracture spacing (Miller, 1979) and fracture density (Chilés, 1988) can be used for a geostatistical evaluation. While variograms of fracture spacing use the number of fracture intersections along a scan-line and assign the spacing to these numbers, variograms of fracture density (Figure 8.4) assign the fracture density (number of intersections per segment) to multiples of these segments. Either variogram should show a pure nugget effect for randomly distributed fractures at the scale of the assessment.

8.2.5 Fracture aperture

Geometric fracture apertures might be collected during scan-line measurements to derive a theoretical aperture distribution. While geometric fracture apertures frequently follow log-normal distributions, the mean value should be adjusted using an effective hydraulic aperture derived from pumping or tracer tests.

8.3 STOCHASTIC FRACTURE NETWORKS

8.3.1 Layer technique

Using the statistical input data for the specialised stochastic fracture network generator KLUFTI described above, the Water Systems Modelling Software SPRING (2012) allows the integration of fracture-specific flow (cubic law) and (advective and diffusive) transport processes into numerical finite-element models. Finite-element models require subdivision of the model domain into discrete elements, which is especially challenging for a fracture network of thousands of fractures linked to each other. Classical algorithms of mesh generation like the advancing front or Delaunay triangulation method are not able to resolve the problem of three-dimensional mesh generation for large-fracture networks. To overcome this problem, SPRING uses the layer technique (Gambolatti and Pini, 1986).

In general, two-dimensional elements are generated first on the surface of the model domain similar to a horizontal model. Then the surface nodes are projected in the vertical direction following user-defined layer boundaries and the nodes for the deeper layers are generated. Modelling of the volume elements (hexahedrons and pentahedrons) is performed layer by layer and variable numbers of nodes in the vertical direction allow layers to end (terminate). Since all surface nodes are projected in the vertical direction, they share x- and y-coordinates, and a single z coordinate defines the position of the interior nodes in space. For discrete fracture network models, three-dimensional rock-matrix elements (hexahedrons and pentahedrons) are linked by their nodes to two-dimensional (triangular and quadrilateral) fracture elements.

8.3.2 Mesh generation

The mesh generation for discrete fracture models starts with the generation of nodes within the fracture planes. Geometrical information like the fracture edge and the intersection lines with other fractures is used to start the generation of nodes in the fracture plane. High flow velocities occur in particular along the intersection lines of fractures and the mesh density has to account for these flow velocities. A surface-meshing technique is therefore used to generate a mesh that is dense at the intersection line but coarse away from it. The same approach is applied for the direction perpendicular to the fracture plane to enable an accurate representation of the hydraulic and diffusive coupling of flow and transport into the rock matrix. Rows of nodes are generated in logarithmic distances parallel to the intersection lines (Figure 8.5).

The remaining area is filled with a raster of nodes chosen by the user. These nodes are then projected to the surface and augmented if the node density is below a user-defined threshold. Surface nodes with a user-defined minimum distance are merged in the geometric strong point to reduce the number of nodes and to improve CPU usage and then re-projected as raster nodes towards layer boundaries. The original fracture planes are then adopted to the raster nodes (potentially causing a minimal distortion of the fracture plane) and volume elements are generated for each layer.

The generation of nodes in 3D demands a minimum number of element layers. A routine checks for each surface node if a fracture occurs in underlying element layers and redefines the layer boundary along the nodes of the fracture plane. If multiple fractures occur within a layer, the layer number is simply raised by the number of additional fractures. If the number of layers resulting from fractures underneath neighbouring nodes is different, the surplus layers end between these nodes (Figure 8.6). If there is no fracture or only one fracture per layer underneath a surface node, the number of element layers corresponds to the user-defined maximum.

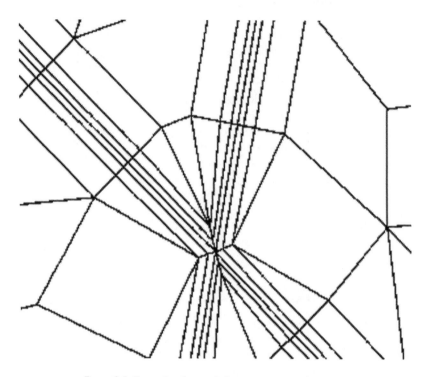

Figure 8.5 Example of a mesh for two crossing fractures.

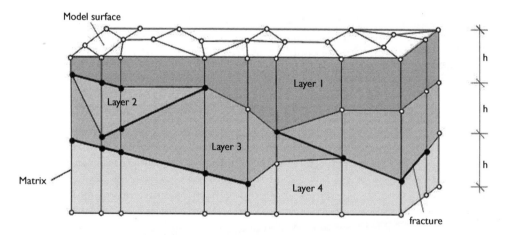

Figure 8.6 Example of variable element layers.

Any number of horizontally trending lithological layers may be generated and fracture orientations and distributions may be individually specified for each layer. Depending on the type of fracture termination, single-, intra- and multilayer fractures can be generated. Additionally SPRING allows the selection of only connected fractures to represent the dominant flow paths and limit CPU time.

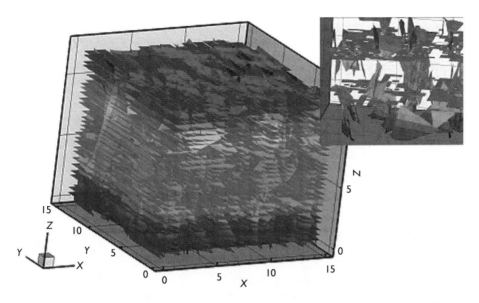

Figure 8.7 Example of a discrete fracture network with channelling effects due to variable fracture apertures.

8.3.3   Examples of discrete fracture network models

Figure 8.7 shows an example of a stochastically generated discrete fracture network within deterministically defined (core logging) bedding planes. Several multi-layer fractures are easily identifiable on the left-hand side. The generated network for a 15 m × 15 m × 15 m block contains 7 600 fractures made up of 254 420 two-dimensional fracture elements.

Constant-head boundaries were applied to opposite sides of the discrete fracture network model (gradient 0.067). Figure 8.8 shows the heterogeneity of flow processes in secondary aquifers, with discrete interconnected fractures dominating the bulk flow across the model domain.

A visualisation of flow velocities within an interconnected fracture network itself (Figure 8.9) highlights the heterogeneity of groundwater flow within a single fracture plane due to variable apertures. The flow within the fractures is generally dominated by a few 'braided' and often intertwined flow channels.

8.4   SIMULATION OF MINE FLOODING

8.4.1   Box-model

Dewatering of mines via abstraction boreholes or capturing of fissure inflows entering the mine voids and pumping to surface drains in the surrounding aquifer will change the aquifer over time from one that is saturated to one that is unsaturated. During active mining operations the horizontal and vertical mine voids are therefore seepage faces and the development of a mine can easily be described numerically using seepage

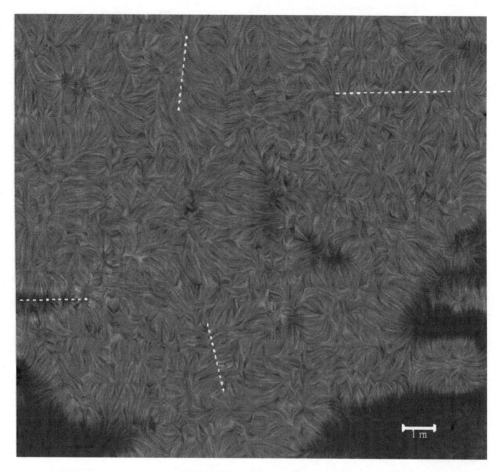

Figure 8.8 Visualisation of flow velocities using the Line Integral Convolution (LIC) method across a vertical profile oblique to the flow direction (selected fractures indicated by hatched lines).

boundary conditions. During flooding, the mine voids can change from a seepage to a recharge face depending on the potential head gradient between the voids and the aquifer. Depending on the layout of a mine, parts might be flooded while others remain drained, resulting in simultaneous water inflows and outflows depending on the prevailing gradient. This gradient changes continuously during the flooding process depending on the water-level rise within the mine and the changing flow regime in the surrounding aquifers.

The simulation of mine flooding with the modelling code SPRING uses a general mine-boundary condition integrated in a box-model based on the features of potential head and seepage-face boundary conditions. Instead of attempting to include discrete mine openings within a regional model, the mine is represented with its major geometrical characteristics relevant for the regional groundwater flow (Figure 8.10). Within a regional flow model any number of cells can be defined as three-dimensional

Figure 8.9 Visualisation of fracture flow velocities using the LIC.

mine compartments, representing, for example, separate shaft or entire mining areas. These compartments are separate water-balance volumes with the same potential head acting as boundary conditions for the neighbouring (groundwater) model cells. Each mine compartment is defined in the model as a completely drained, unsaturated zone, for which the stage-volume relationship (head/elevation dependent water storage)

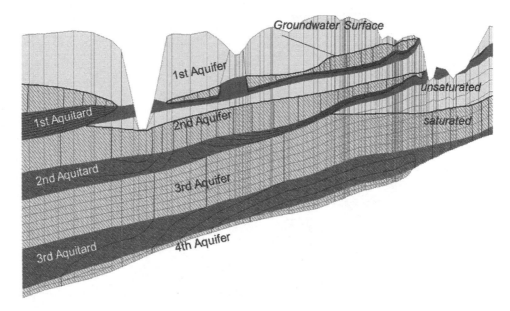

Figure 8.10 Example of potential heads and free groundwater surfaces in a multi-layer aquifer system.

and active mine-water management interventions (e.g. pumping rates over time) are required (Figure 8.11). The interaction with the surrounding groundwater body is then calculated by the model depending on the prevailing flooding level.

At the beginning of the flooding simulation, a free seepage-face boundary condition is assigned to all nodes of the mine compartment. Such a boundary condition, usually used to calculate the seepage surface through a dam, allows calculation of the (unsaturated) seepage volume into a mine void as long as the potential head of a single model node is below its geodetic height. Once the potential head exceeds the geodetic height of the node, the boundary condition changes to a potential head boundary with the head $h(V_i)$ for this node as well as for all previously flooded nodes.

The water balance for each delineated mine compartment is calculated for each time step of the model. The water stored in a mine compartment is calculated for each time step $t = t_i$ as the balance of all volume fluxes from the time $t = 0$ (begin of flooding) to the current time step. The water stored in a mine at a time t_i is therefore the sum of water fluxes during the current time step $t = t_i - t_{i-1}$ and the total volume stored up to the preceding time step $t = t_{i-1}$. Using the stage-volume relationship (Figure 8.12), the new potential head $h(V_i)$ within the mine compartment is then calculated. The approach ensures conservation of mass as the volume of water extracted from the model via the free seepage face is fully transferred to the recalculated potential head boundary. Based on the recalculated potential head, the boundary conditions for the mine cells are updated and the hydraulic heads in the regional flow field calculated, resulting in flow from or to the mine compartment. If required, leakage instead of potential head boundaries can be assigned to limit fluxes between flooded

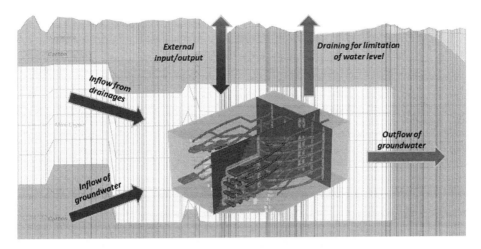

Figure 8.11 Conceptual model for the simulation of mine flooding.

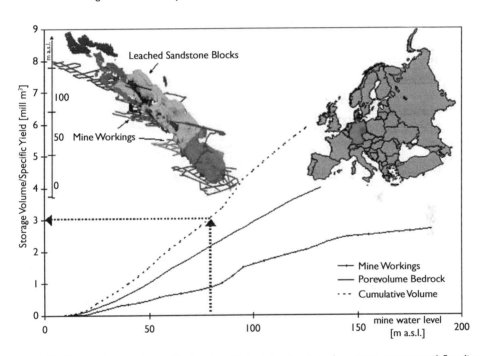

Figure 8.12 Stage-volume relationship for the Königstein mine based on mine surveys and flooding experiments (inset: 3D-visualisation of mine workings (left) and location of Königstein mine (right)).

mine voids and the aquifer as a result of, for instance, previous fissure grouting to stop water inflows.

If all nodes of a mine compartment are assigned potential head boundaries, the mine is considered flooded and the boundary conditions removed. The mine then becomes a completely saturated part of the aquifer and flow calculations (with unique 'aquifer' properties). If the potential head drops again, the nodes of the mine voids are reactivated and as for the flooding process set to free seepage potential head boundaries. If the flooding of a mine compartment is limited to a specified constant (e.g. environmental critical level) or time-dependant elevation, the model calculates the required (time dependant) pumping volumes to maintain the level(s).

8.5 CONCLUSION

With the advancement of personal computer capacities, the modelling of discrete fracture network models becomes a viable tool for quantitative risk assessments in fractured-rock environments. The Water Systems Modelling Software SPRING has the capacity to stochastically generate fracture network models and to incorporate them into numerical finite-element models. SPRING uses the layer technique for three-dimensional mesh generation, where three-dimensional rock matrix elements are linked by their nodes to two-dimensional fracture elements. To account for the highly variable flow and transport processes in a fully coupled fracture-matrix model, surface-meshing techniques with logarithmically spaced nodes parallel to fracture intersections and perpendicular to fracture planes are used. Built-in software routines optimise the number of notes and layers required to represent complex fracture networks.

Modelling the flooding of an underground mine in a regional flow model faces similar challenges regarding appropriate parameterisation of the mine and discretisation of the mine openings. The box-model approach for mine compartments allows for numerically stable modelling of underground mine-flooding processes within a regional groundwater-flow model, whilst reducing the necessary discretisation effort. The development of the mine-water level as well as changes in regional flow conditions can be predicted in one model run for the dewatering and flooding of the underground mine. A coherent water balance of the whole and/or separate mine compartments of an underground mine is calculated for each time step based on a water-storage characteristics of the underground mine.

While the presented methods enable flow and transport calculations of fractured saturated and unsaturated rocks as well as the assessment of mining impacts in fractured aquifers, the required data-collection efforts and CPU times are substantial.

REFERENCES

Adam, J.F. (1989) Methoden und Algorithmen zur Verwaltung und Analyse axialer 3-D-Richtungsdaten und ihrer Belegungsdichte. U-Thesis, University of Göttingen, Göttingen, Germany.

Black, J.H. (1994) Hydrogeology of fractured rocks – A question of uncertainty about geometry. *Applied Hydrogeology* 2, 56–70.

Chilés, J.P. (1988) Fractal and geostatistical methods for modeling of a fracture network. *Mathematical Geology* 20, 631–654.

Chilés, J.P. & de Marsily, G. (1993) Stochastic models of fracture systems and their use in flow and transport modeling. In: Bear, J., Tsang, C.F. & de Marsily, G. (eds.) *Flow and containment transport in fractured rock*. San Diego, CA, Academic Press, 169–236.

Gambolatti, G. & Pini, G. (1986) A 3-D finite element conjugate gradient model of subsurface flow with automatic mesh generation. *Advances in Water Resources* 9, 1486–1493.

Kulatilake, P.H., Wathugala, D.N. & Stephansson, O. (1993) Joint network modeling with a validation exercise in Stripa Mine, Sweden. *International Journal of Rock Mechanics and Mining Sciences & Geomechanics Abstracts* 30, 503–526.

Miller, S.M. (1979) Geostatistical analysis for evaluating spatial dependence of fracture set characteristics. In: O'Neil, T.J. (ed.) *Proceedings of the 16th APCOM Symposium*. Tucson, Arizona, New York, NY, SME-AIME. pp. 537–545.

Priest, S.D. (1993) The collection and analysis of discontinuity orientation data for engineering design, with examples. In: Hudson, J.A. (ed.) *Comprehensive rock engineering, principals, practice and projects*. Volume 3. Oxford, UK, Pergamon Press.

Priest, S.D. & Hudson, J.A. (1981) Estimation of discontinuity spacing and trace length using scanline surveys. *International Journal of Rock Mechanics and Mining Sciences & Geomechanics Abstracts* 22, 135–148.

SPRING (2012) User's Manual Version 4.1, Delta-H Water Systems Modelling, Pretoria, South Africa.

Terzaghi, R.D. (1965) Sources of error in joint surveys. *Geotechnique* 15, 287–304.

Chapter 9

Assessing uncertainties in surface-water and groundwater interaction modelling – a case study from South Africa using the Pitman model

J.L. Tanner & D.A. Hughes
Institute for Water Research, Rhodes University, Grahamstown, South Africa

ABSTRACT

Understanding surface-water and groundwater interactions plays a major role in the successful implementation of integrated water-resource management. Challenges have arisen due to the lack of observed data, differences in surface-water and groundwater catchment boundaries and because different methods have traditionally been used for assessing surface water and groundwater. The result is a great deal of uncertainty when modelling interactions. One approach to integration includes the addition of more explicit groundwater algorithms into the commonly used Pitman hydrological model. This study uses the model to explore the main sources of uncertainty identified in the headwaters of the Breede River located in the Western Cape Province, South Africa. The results are presented within an uncertainty framework facilitating a range of possible process descriptions, rather than one hydrological time series. Despite attempting to more clearly identify the dominant sources of uncertainty in the catchment, the range of uncertainty did not vary appreciably. The complexity of the processes in the catchment seems to have resulted in an unresolved amount of uncertainty from a number of sources. While some uncertainties can be reduced (notably water use and recharge), this can only be achieved through substantial investments of human and financial resources. The results suggest large uncertainty in natural and impacted low flows, which has serious implications for water management related to environmental flow legislation.

9.1 INTRODUCTION

It has emerged that understanding surface-water and groundwater interactions plays a large role in the successful implementation of Integrated Water-Resource Management (IWRM). A better conceptual understanding and quantitative description of interactions between surface water and groundwater is needed, and efforts to link surface-water and groundwater estimation methods are now being made. Challenges have arisen due to the lack of observed data, differences in surface-water and groundwater catchment boundaries and the fact that different methods have traditionally been used for assessing surface water and groundwater. The result is a great deal of uncertainty when modelling surface-water and groundwater interactions.

One approach to surface-water and groundwater integration is the addition of more explicit (and conceptually appropriate) groundwater algorithms into the commonly used Pitman hydrological model (Hughes, 2004) for the purpose of incorporating surface-water and groundwater interactions and the impacts of groundwater abstraction. The model is designed to complement detailed surface-water and groundwater models which are appropriate for detailed assessments of each resource, but need to be integrated when modelling catchment-scale interaction variables for the purpose of water-resource management. The model is seen as a first step towards understanding the temporal and spatial patterns of surface-water and groundwater interactions and aims to establish at least first-order estimates of the rate and volume of exchange between groundwater and surface water.

Historically, the reliability of model results has been evaluated using observed data (where available). More recent approaches to the application of hydrological and water-resource-estimation models have focused on the explicit quantification of the uncertainties in the outputs (Hughes and Mantel, 2010a). The science of hydrology is currently experiencing a shift from methods that focus on the identification of a single best model towards methods that attempt to reduce the uncertainty in the predictions of all possible models using various types of ensemble methods. Instead of focusing on 'optimisation', the focus is changing towards one of model 'consistency' (Kapangaziwiri and Hughes, 2009). Many hydrology and hydraulic researchers now see uncertainty analysis as an important part of good scientific practice and an extensive range of papers is available on uncertainty-estimation methods and applications (Pappenberger and Beven, 2006). In this study, a number of model output ensembles of possible process descriptions for a basin are produced rather than one hydrological time series as has previously been the norm. Beven (2001) states that it is impossible to prove a constructed water balance without allowing significant uncertainty in every input and output term. Similarly, Kapangaziwiri and Hughes (2009) point out that the notion of an optimal parameter set is considered both unwise and incorrect, especially when considered against input forcing data uncertainty, model structures and incomplete and limited process understanding.

There will always be higher levels of risk associated with the management of water resources in South Africa due to the high degree of both spatial and temporal variability in available resources. The initial focus of any attempt to reduce uncertainty in the outputs of models should be the identification of the main sources of the uncertainty. A WRC project (Hughes and Kapangaziwiri, 2010) completed in 2010 identified the main sources of uncertainty associated with hydrological models. These included the input hydrometeorological data (typically rainfall and evaporation demand in rainfall-runoff models), the parameters used to represent the processes in a basin, model structure (including spatial and temporal-scale issues), and water use or return flows. In many circumstances it is difficult to fully understand the sources of uncertainty and therefore they cannot be properly quantified.

Therefore it can be argued that a substantial contribution to uncertainty reduction can be made through a better quantitative understanding of the different sources of uncertainty and their relative contribution to total uncertainty (Hughes, 2012). In this study, the modified Pitman model (Hughes, 2004) is used within an uncertainty framework and contributes to previous work which examined different areas of uncertainty in more detail. These include the uncertainties associated with physically based *a priori* parameter-estimation methods (Kapangaziwiri and Hughes, 2008), estimates of the impacts of farm dams (Hughes and Mantel, 2010a), and different rainfall datasets (Hughes and Mantel, 2010b). These studies illustrated approaches that can be used to estimate uncertainties in simulations of natural hydrology and the impacts of water-resource developments and were able to isolate the dominant sources of uncertainty in the respective example catchments. In this study, a similar methodology is followed with the uncertainties associated with some of the groundwater parameters of the model examined (recharge, inter-basin transfers of groundwater and groundwater use).

9.2 THE MODEL AND UNCERTAINTY ESTIMATES

The Pitman model has been widely applied in the Southern Africa region. The reader is referred to previous publications (Pitman, 1973; Hughes, 2004; Hughes *et al.*, 2007) for the detailed structure of the model. The model comprises three conceptual storages (interception, soil moisture and groundwater) and simulates infiltration-excess flow, saturation-excess flow, direct overland flow and groundwater flow. It is a conceptual, semi-distributed, monthly rainfall-runoff model that uses monthly rainfall data and monthly estimates of evapotranspiration as input. The groundwater version of the Pitman model (Hughes, 2004), used in this study, has explicit groundwater routines and is quite heavily parameterised. While the model has been demonstrated to be applicable to most areas, there are some structural uncertainties associated with the spatial and temporal scales used. However, these are difficult to isolate from the uncertainties associated with establishing appropriate parameter sets.

This study utilises a model application framework that explicitly deals with the issue of uncertainty (Kapangaziwiri and Hughes, 2009) as well as parameter-estimation routines for the Pitman model that have been modified to fit into this framework (Kapangaziwiri and Hughes, 2008). These routines attempt to translate uncertainty in the available physical basin property data into uncertainty in the resulting estimates of parameter values. Details of the uncertainty framework can be found in Kapangaziwiri and Hughes (2009) with the major components of the framework including: (1) prior parameter distributions; (2) parameter sampling; (3) the production of output ensembles; and (4) the extraction of behavioural/consistent samples using regional constraints. The output file produced by the model consists of a text file of all parameter sets (usually 10 000), the constraint data (mean monthly runoff, mean monthly recharge, slope of the flow duration curve, the 90th, 50th and the 10th percentiles) and five objective functions (the coefficient of efficiency for the normal (CE N), natural logarithm transformed (CE L) and inverse values (CE 1/d) and the mean monthly runoff error for both normal (% Diff N) and natural logarithm transformed values (% Diff L)).

9.2.1 Unknown uncertainties

9.2.1.1 Observed streamflow data

The problem of poorly gauged and un-gauged basins has undermined the efficacy and reliability of making hydrological predictions in many regions of the world, especially in Southern Africa. Even where historical records of data are available, there is often the problem of unaccounted for and poorly documented upstream human interferences, and it is frequently difficult to isolate the natural hydrology of the basins (Hughes and Mantel, 2010a). In this study, the observed data for the example basin (gauge H1H003 located at the outlet of quaternary catchment H10C) (see Figure 9.1) is substantially impacted by upstream interferences. In addition, while the simulated flows represent stationary development conditions associated with the fixed parameter values, the observed flow data are expected to be non-stationary, reflecting the history of water-resource development (Hughes and Mantel, 2010b).

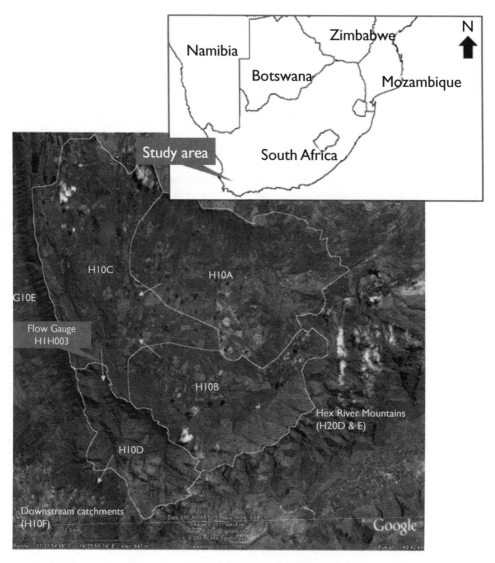

Figure 9.1 Quaternary catchments H10A–D with arrows representing surface-water flow (figure extracted from Google Earth – Eye elevation: 40.92 km).

9.2.1.2 Water-use uncertainty

This research examines both surface-water and groundwater use and attempts are made to identify which type of water use contributes to the bulk of the water use in the example catchment. Hughes and Mantel (2010a) discussed the details of uncertainty estimation for the parameters affecting the simulation of farm-dam effects, and quantified the volume of abstractions based on the irrigated area in the basin. They identified more than 350 farm dams in the area of the present study, using information generated by both the Chief Directorate of Surveys and Land Information and Google Earth images. The present study has incorporated groundwater use as a percentage of the total water use quantified in the Hughes and Mantel (2010a) study. Groundwater-use data from GRA II (DWAF, 2005) and the Berg Water Availability Assessment Study (WAAS) (DWAF, 2007) were used.

9.2.1.3 Rainfall uncertainty

There are many different ways in which a sub-basin average rainfall time series can be generated from the available rain gauge data. The uncertainties are expected to be associated with the density and representativeness of the gauging network, the reliability of the available data and the degree of spatial variability of the real rainfall (Hughes and Mantel, 2010b). In this study, the WR2005 data (Middleton and Bailey, 2008) were used as it is the more recent dataset, although one simulation was carried out using the WR90 data (Midgley *et al.*, 1994) for comparative purposes.

9.2.2 Remaining uncertainties

9.2.2.1 Parameter uncertainty

There is very little explicit information available to quantify many of the processes involved in surface-water and groundwater interactions, and therefore, the relevant model parameters. For example, groundwater recharge is generally assumed to either leave the immediate surface-water catchment as sub-surface transfers to other surface-water catchments, contribute to streamflow within the surface-water catchment, be lost to evapotranspiration (in the riparian margins of the channel perhaps), or be abstracted. It is not straightforward to trace the path of recharge water on a scale that might be useful for water-resource management.

For this work, an uncertain parameter set was generated using the set of parameter-estimation equations developed by Kapangaziwiri and Hughes (2008). These are based on the use of physical basin property data, including topography, soils, vegetation and geological information. The basic tenet of the parameter-estimation procedures is that physical attributes play a major role in the hydrological conditioning of catchments and that it is therefore possible to use these measurable properties to directly quantify model parameters. Winter (2000) argues that many seemingly diverse landscapes have some features in common, and it is these commonalities that need to be identified. Only by evaluating landscapes from a common conceptual framework can processes common to some or all landscapes be distinguished from processes unique to particular landscapes.

Although details of the parameter-estimation equations can be found in Kapangaziwiri and Hughes (2008) and Kapangaziwiri and Hughes (2009), the basic methodology is summarised below. The incorporation of uncertainty into the parameter-estimation procedures is based on definitions of suitable frequency distributions for the input physical property data. The raw basin characteristic data, together with measures of uncertainty, are input into the estimation procedures as primary basin data that are used to calculate the secondary data. The primary data are usually point- or small-scale measurements. Since the model operates at the basin scale, its parameters need to be estimated as averages capable of adequately describing relevant basin-wide processes. Therefore secondary data are calculated from the primary variables in an attempt to estimate the required basin averages of the physical attributes data and are then used to estimate the parameters. The calculation of each basin-scale secondary variable is based on sampling from all the relevant primary variables on the basis of an assumed normal frequency distribution defined by specifying the mean and standard deviation values associated with each primary variable. Different sample sizes of the primary variables are taken depending on proportion of catchment area covered. A Monte Carlo sampling approach is used to randomly sample from the probability distributions and the results are used to determine the mean, standard deviation and skewness of the secondary variables. A similar procedure is used for the final estimation of parameters from the secondary data.

9.2.2.2   Recharge uncertainty

There is not enough information about recharge processes and their temporal variability at the catchment scale. The GRA II estimates are very uncertain and give no information about temporal variability. Without accurate estimates of recharge patterns, the other groundwater components of the model will remain highly uncertain. The difficulty in assigning recharge values to a basin has resulted in appreciable variation within the different estimates for the same locality. The GRA II estimates of recharge are frequently used in the modified Pitman model with an estimate between the lower and middle value deemed to be appropriate in most basins. In this study, the recharge was increased to match the Berg WAAS (DWAF, 2007) recharge estimation which was similar to or greater than the highest GRA II estimate for the quaternary catchments in the example basin.

9.2.2.3   Regional groundwater flow uncertainty

It is known that surface-water and groundwater catchment boundaries do not always coincide, and that this issue has been neglected in surface hydrology models that have recently included groundwater components. The authors acknowledge that the current version of the model is not appropriate in situations where groundwater and surface-water divides of a basin are not the same, although future model updates could allow for scenarios where regional groundwater flows dominate drainage processes. In this study, a conceptual model of the example catchment indicated the possibility of the existence of regional groundwater flows. In an attempt to simulate this process, the groundwater slope of the model was increased causing a high volume of downstream groundwater outflow. Although the outflow is moved into a downstream catchment, for the purposes of this study it was seen to replicate the regional flow by removing the outflow from the local catchment water balance. However, there is very little direct information on the sub-surface routing of groundwater flow. Before this type of process can be satisfactorily included in models, some guidelines are required about how the relevant parameters would be quantified.

9.2.3   Combining uncertainties

The first step in the approach adopted in this study was to generate 10 000 ensembles of natural hydrology using defined uncertainty (from the parameter-estimation tool) in the natural runoff parameters of the model. The second step was to increase the recharge from the lower estimate of the GRA II database, to the recharge given in the Berg WAAS (DWAF, 2007) and to re-run the 10 000 ensembles. The third step was to increase the groundwater slope until a higher proportion of groundwater outflow occurred which represented the sub-surface inter-basin transfer of water. The Berg WAAS (DWAF, 2007) gives an estimate of the amount of groundwater lost through this process. This figure was increased as the value reported by DWAF (2007) was too small to have any substantial effect on the water balance. The 10 000 ensembles were then re-run. Using the water-use data from GRA II and the Berg WAAS, a proportion of the total water use (Hughes and Mantel, 2010b) was assigned to groundwater use before the 10 000 ensembles were run for the fourth and fifth time (with surface-water use, then with groundwater use incorporated). The resulting uncertainty parameter set which includes all the stages of the assessment is given in Table 9.1. The output from each model run included three time series based on ranking (ascending order) the 10 000 simulated flows for each month of the time series. The 'lower' time series represents the 5th percentile of the ranked values, while the 'central' and 'upper' time series represent the 50th and 95th percentiles, respectively. The lower and upper time series therefore represent the bands covering 90% of all simulated flows.

9.2.4   Model sensitivity analysis

A regionalised sensitivity analysis (Wagener et al., 2002; Kapangaziwiri, 2010) was also applied to evaluate the parameter sensitivity within the catchments. Although no calibration was carried out in this study, a sensitivity analysis can identify the parameters and/or parameter combinations that lead to non-behavioural ensembles. The results of the regionalised sensitivity analysis in this study are based on the measure of distribution of the model response that resulted from the 10 000 Monte Carlo input-parameter groups sampled. The output ensembles are ranked on the basis of the assessment criteria sorted into five equal groups, then normalised cumulative distribution curves are plotted (Y-axis) for each parameter (X-axis). The sensitivity of the parameter is measured by the degree of divergence of the cumulative curves, i.e. the wide separation of the curves indicates that the parameter is very sensitive based on the assessment criteria considered. Two categories of assessment criteria used in this study are the flow metric (i.e., Mean Monthly Flow (MMF), Mean Monthly Recharge (MMR), slope of the flow-duration curve, the 10th, 50th and 90th percentiles of the cumulative frequency distribution of flows) and the objective functions (Tshimanga et al., 2011).

Table 9.1 The physically based parameter estimates used in the simulations (distribution type 1: normal distribution; 2: log-normal distribution; 3: uniform distribution).

Parameters	Mean	Standard deviation	Skewness	Distribution type	Minimum value	Maximum value
Rain distribution factor	1.28	0	0	0	0	0
Proportion of impervious area AI	0	0	0	0	0	0
Summer intercept cap.(Veg1) PI1s	0.6	0.067	0.155	1	0	5
Winter intercept cap.(Veg1) PI1w	0.601	0.067	0.155	1	0	5
Summer intercept cap.(Veg2) PI2s	2.348	0.013	0.009	1	0	5
Winter intercept cap.(Veg2) PI2w	2.348	0.013	−0.062	1	0	5
% Area of Veg2 AFOR	0	0	0	0	0	0
Veg2/Veg1 pot. evap. ratio FF	1.4	0	0	0	0	0
Power of Veg recession curve	0	0	0	0	0	0
Annual pan evaporation (mm) PEVAP	1650	0	0	0	0	0
Summer min.abs.rate (mm/mth) ZMINs	93.6	12	−5.192	1	0	200
Winter min.abs.rate (mm/mth) ZMINw	93.6	12	−5.192	1	0	200
Mean abs.rate (mm/mth) ZAVE	774.178	0	0	0	0	0
Maximum abs.rate (mm/mth) ZMAX	999.6	38.136	1.375	1	0	5 000
Maximum storage capacity ST	166.492	23.261	0.066	1	10	5 000
No recharge below storage SL	0	0	0	0	0	0
Power: storage-runoff curve POW	1.93	0.1	−0.65	1	1	10
Runoff rate at ST (mm/mth) FT	36.554	17.428	0.607	1	0	1 000
Max. recharge rate (mm/mth) GW	23	2	0.563	1	0	1 000
Evaporation-storage coefficient R	0.5	1	0	3	0.7	1
Surface runoff time lag (months) TL	0.25	0	0	0	0	0
Channel routing coeff. CL	0	0	0	0	0	0
Irrig. area (km²) AIRR	0	0	0	0	0	0
Irrig. return flow fraction IWR	0	0	0	0	0	0
Effective rainfall fraction	0	0	0	0	0	0
Non-irrig. direct demand (Ml/yr)	0	0	0	0	0	0
Maximum dam storage (Ml)	23 907	1.734	0	1	0	30 650
% Catchment area above dams	45.8	0	0	3	45.8	68.6
A in area volume relationship	6.5	0	0	3	6.5	7.9
B in area volume relationship	0.8	0	0	0	0	0
Irrig. area from dams (km²)	16	8.4	0	1	10	90
Channel loss TLG max (mm)	0	0	0	0	0	0
Power: storage-recharge curve GPOW	3	0	0	0	0	0
Drainage density	0.4	0	0	0	0	0
Transmissivity (m²/d)	50	10	0	1	1	500
Storativity	0.001	0.001	0	1	0.001	0.8
Initial GW drainage slope	0.05	0	0	0	0	0
Rest water level (m below surface)	25	0	0	0	0	0
Riparian strip factor	0.2	0.02	0	1	0	0.6
GW abstraction (upper slopes-Ml/yr)	3 023.5	0	0	0	0	0
GW abstraction (lower slopes-Ml/yr)	12 094	0	0	0	0	0

9.3   EXAMPLE SUB-BASIN

The H10A–C group of sub-catchments are located in the headwaters of the Breede River in the winter-rainfall region of the Western Cape Province (Figure 9.1). The area forms part of the Cape Fold Belt and the landscape consists of steep mountain slopes and flat valley floors. Table Mountain Sandstone (TMS) is exposed on the top and flanks of the anticlinal folded mountains and is confined in the synclinal valleys by shale and mudstones of the Bokkeveld Group (DWAF, 2008). The dominant land-use activity is deciduous fruit cultivation supported by irrigation from a large number (>350) of small farm dams (Hughes and Mantel, 2010b). The area experiences a typical Mediterranean climate with moderate temperatures and winter rainfall. As a result of the orographic control of rainfall, the Mean Annual Precipitation (MAP) varies significantly across the study area (approximately 500 mm/yr in the valley bottom areas to over 1 000 mm/yr in the mountains), while mean annual potential evaporation is between 1 600 mm/yr to 1 700 mm/yr. The flow regime is highly variable, partly as a consequence of the rainfall seasonality (coincidence of low rainfall and high evaporation months). The AGIS (2007) land-type data, which were used to define the sub-basin physical properties, indicate a large spatial variation in soil characteristics that would be expected to lead to substantial uncertainty in the estimation of parameter values (Hughes and Mantel, 2010b).

The two major formations making up the TMS are the Peninsula and Nardouw Formations, and are overlain by the Bokkeveld Group. The fold system is the main structural element forming natural boundaries to groundwater and surface-water flow (DWAF, 2008). A conceptual model of water flow was provided by the Berg WAAS and is summarised below. Surface-water flow drains from the bounding mountain ranges into the valley and out through Mitchell's Pass in the south-western corner (H10D). The groundwater flow is more complex with the three main aquifers in the catchment following different flow paths. The Bokkeveld aquifer and Nardouw TMS aquifer both have their recharge zones and discharge zones located within the catchment. However, although the Peninsula TMS aquifer recharges in the catchment, it discharges into the Hex River catchment system to the east through deep groundwater flow (DWAF, 2008).

9.4   RESULTS AND DISCUSSION

The results of the comparisons are presented as Flow-Duration Curves (FDCs) in Figure 9.2. Based on the relative spread of the simulated ensembles around the central value, there is a great deal more uncertainty in the simulation of natural low flows (90% exceedance) than for moderate to high flows (50% and 10% exceedance). This is largely a consequence of the uncertainty in the drainage and groundwater recharge and discharge parameters of the model.

Increasing the recharge to match the values given in the Berg WAAS (DWAF, 2007) did not have any appreciable effect on the magnitude of uncertainty. Both estimates of uncertainty (GRA II – Figure 9.2a and Berg WAAS – Figure 9.2b) remain highly uncertain due to a lack of data and the variability within this catchment.

The conceptual model given in the Berg WAAS indicates that groundwater recharge into the folded Table Mountain Sandstone aquifers could be moving out of the catchment through regional groundwater flow and discharge into the adjacent

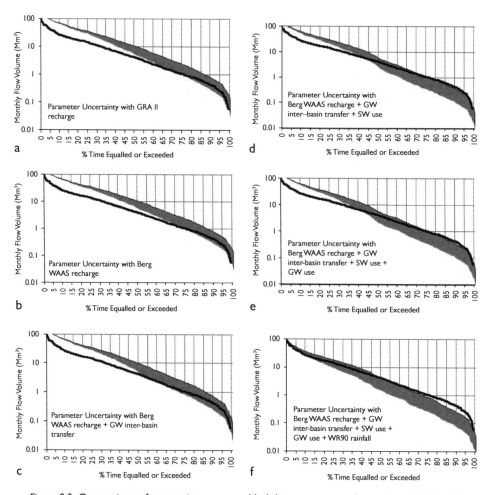

Figure 9.2 Comparison of uncertainty sources (dark line represents observed patched data).

Hex River basin. The volume of recharge into the TMS aquifer is given in the Berg WAAS study and was simulated as groundwater outflow (Figure 9.2c). There is however, very little information available on this process, and hence, the inclusion of this process in the model is highly uncertain. In addition, simulating water use as either surface water (Figure 9.2d) or groundwater (Figure 9.2e) did not have a significant impact on the uncertainty bounds, so the simulations were unable to confirm whether the water use is predominantly from surface water or groundwater.

The rainfall figures used in this study were from the WR2005 (Middleton and Bailey, 2008) data. The last simulation, however, was run with WR90 (Midgley *et al.*, 1994) rainfall data (Figure 9.2f). It was clear that different assumptions had been made about the orographic effects (which are large in this catchment) when comparing the rainfall data from the two regional studies (WR90 and WR2005). The WR90 study assumed a mean annual rainfall across the whole of H10 of 622 mm/yr compared with 863 mm/yr for the WR2005 study (Middleton and Bailey, 2008) study.

The observed data used in this study also contain a number of uncertainties which further complicate the interpretation of the simulation results. The observed data for flow gauge H1H003, located at the outlet of quaternary catchment H10C, records flows from 1993 to the present. These data were patched during the WR90 (Midgeley *et al.*, 1994) study; however, a number of uncertainties in the data remain. A similar patching exercise for H1H003 was carried out by the authors, and it was found that some high flows were above the gauge rating curve limit. Therefore, no information on the volume of flow during those periods exists within the records. In addition, the low flows are partly impacted by wastewater return flows and storm runoff from the urban area of Ceres (with a population of about 40 000 and located close to the outlet of H10C), which would be reflected in the observed records, but have not been accounted for in the model set-up. A rough calculation carried out to determine an approximate volume of return flows (approximate average water use per person = 15 m³/month ∗ % of the water used likely to contribute to return flows (60%) ∗ approximate population of Ceres (40 000)) gives a total of about 4.32 Mm³/yr for Ceres. This volume is likely to have an impact on the low flows in the river and could partly account for the under-simulation of the model when abstractions are accounted for (Figure 9.2d–f).

The sensitivity tests carried out highlighted the influential and non-influential parameters for the basin. Given the large number of the sensitivity analysis results, only samples of plots are shown to illustrate the trend of parameter sensitivity in the basin (see Table 9.2 and Figure 9.3). There is little doubt that some of the non-influential parameters could be a result of an inappropriate combination of the parameters.

Table 9.2 Sensitivity analysis results for selected parameters. H represents a high sensitivity, M represents a moderate sensitivity and L represents a low sensitivity.

Parameter	Flow metrics					Objective functions				
	Mean monthly flow	Slope of FDC	Q10	Q50	Q90	CE N	CE L	CE I/data	% Diff N	% Diff L
H10A										
ST	H	H		M						
POW										
FT	H	M	H	H						
GW	M	L								
T	H	H	M	M						
H10B										
ST	L	M		L						
POW	L		L	M						
FT	H	H	H	H						
GW										
T										
H10C										
ST	M	M		M	M	L	H	M		
POW	L		L	M		M	H	L	M	
FT	H	H	H	H	H	H	H	H	H	
GW										
T										

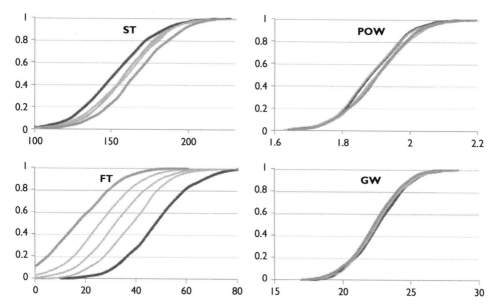

Figure 9.3 Model parameter sensitivity for selected parameters. The black line, light-grey lines and
dark-grey line represent the top 20%, middle 60% and lower 20% of the ensembles, respectively. All of these parameters have been assessed on the mean monthly flow criteria and
represent highly sensitive (FT), moderately sensitive (ST), weakly sensitive (POW) and
insensitive (GW) parameters.

The model simulations were very sensitive to the parameters ST and FT in particular
based on various assessment criteria used in the sensitivity tests. These parameters
represent the soil moisture store and the interflow processes within the model, which
suggests that interflow plays an important role in the Ceres Basin. The dominance of
the Table Mountain Sandstone in the catchment would support this.

9.5 CONCLUSIONS

Despite attempting to more clearly identify the dominant sources of uncertainty, the
complexity of the processes in this catchment has resulted in an unresolved amount
of uncertainty from a number of sources. The alternative recharge values, the incorporation of groundwater use and the inclusion of groundwater inter-basin transfers
all produced a similar quantity of uncertainty in the outputs of the model simulations, but with somewhat different deviations from the observed flow records. The
Ceres basin is a particularly complex basin with a high degree of variability, and
many sources contributing to the overall uncertainty. While in this circumstance it
has proved difficult to isolate the dominant sources of uncertainty, it is still essential
to identify and evaluate these sources as far as possible. An improved understanding
of the sources and magnitude of uncertainty is needed, as well as a reduction of the
uncertainty through improved monitoring, estimation methods and training. Reducing

input rainfall uncertainty requires substantial resources to expand existing monitoring networks. Water-use uncertainties can always be reduced through detailed field surveys; however, these will also require quite large investments of time and money. It is also important to bear in mind the type of data the investigations are designed to collect. The question of data richness depends not only on the amount of data, but also on whether or not the data are directly appropriate for the type of water-resource management and planning decisions that have to be made. This is the difference between data richness from a purely hydrological perspective and from a practical water-resource management perspective (Hughes, 2012). This study further emphasises the importance of obtaining suitable values for the groundwater parameters of the model, and highlights the large differences in available recharge estimates. The major sources of uncertainty are therefore common to any modelling approach and resolving these on the basis of improved field information should be considered a priority. The risks of not accounting for uncertainty in model simulations are increasing as our water resources become more stressed. At the same time, however, it is important to improve the decision-making process and not to make decisions impossible by raising too many issues of uncertainty (Hughes and Kapangaziwiri, 2010).

ACKNOWLEDGEMENTS

Ms Tanner is a Ph.D. student in the Institute for Water Research and is financially supported by the Carnegie Foundation of New York under the auspices of the Sub-Saharan Africa Water Resources Network (SSAWRN) which is part of the Regional Initiative for Science Education (RISE) programme, as well as the German Academic Exchange Program (DAAD) in conjunction with the National Research Foundation (NRF).

REFERENCES

Beven, K.J. (2001) On hypothesis testing in hydrology. *Hydrological Processes* 15, 1655–1657.
DWAF (Department of Water Affairs and Forestry) (2005) *Groundwater Resource Assessment Phase II (GRAII).* Pretoria, South Africa, DWAF.
DWAF (Department of Water Affairs and Forestry) (2007) *The assessment of water availability in the Berg Catchment (WMA 19) by means of water resource related models: Ground water model report, Volume 4 – Regional water balance model.* Prepared by Umvoto Africa (Pty) Ltd in association with Ninham Shand (Pty) Ltd on behalf of the Directorate, National Water Resource Planning. Pretoria, South Africa, DWAF. DWAF Report No. PWMA 19/000/00/0407.
DWAF (Department of Water Affairs and Forestry) (2008) *The assessment of water availability in the Berg Catchment (WMA 19) by means of water resource related models: Groundwater model report, Volume 3 – Regional conceptual model.* Prepared by Umvoto Africa (Pty) Ltd in association with Ninham Shand (Pty) Ltd on behalf of the Directorate, National Water Resource Planning. Pretoria, South Africa, DWAF. DWAF Report No. PWMA 19/000/00/0408.
Hughes, D.A. (2004) Incorporating groundwater recharge and discharge functions into an existing monthly rainfall runoff model. *Hydrological Sciences Journal* 49, 297–311.

Hughes, D.A. (2012) *Implementing uncertainty analysis in water resources assessment and planning*. Deliverable No. 3: First general report on uncertainty reduction. Pretoria, South Africa, Water Research Commission. WRC Project No: K5/2056.

Hughes, D.A. & Kapangaziwiri, E. (2010) *Identification, estimation, quantification and incorporation of risk and uncertainty in water resources management tools in South Africa*. Deliverable No. 11: A report on reducing uncertainty. Pretoria, South Africa, Water Research Commission. WRC Project No: K5/1838.

Hughes, D.A. & Mantel, S.K. (2010a) Estimating the uncertainty in the impacts of small farm dams on stream flow regimes in South Africa. *Hydrological Sciences Journal* 55, 578–592.

Hughes, D.A. & Mantel, S.K. (2010b) Estimating uncertainties in simulations of natural and modified streamflow regimes in South Africa. In: *Proceedings of the 6th World FRIEND Conference, Fez, Morocco, October 2010*. IAHS Publ. 340, 2010.

Hughes, D.A., Parsons, R., & Conrad, J. (2007) *Quantification of the groundwater contribution to baseflow*. Pretoria, South Africa, Water Research Commission. WRC Report No. 1498/1/07.

Kapangaziwiri, E. (2010) Regional application of the Pitman monthly rainfall-runoff model in southern Africa incorporating uncertainty. PhD thesis, Rhodes University, Grahamstown, South Africa. [Online] Available from: http://eprints.ru.ac.za/1777/ [Accessed July 2011].

Kapangaziwiri, E. & Hughes, D.A. (2008) Towards revised physically based parameter estimation methods for the Pitman monthly rainfall-runoff model. *Water SA* 34, 183–192.

Kapangaziwiri, E. & Hughes, D.A. (2009) Assessing uncertainty in the generation of natural hydrology scenarios using the Pitman monthly model. In: *Proceedings of the 14th South African National Hydrology Symposium, Pietermaritzburg, 2009*.

Middleton, B.J. & Bailey, A.K. (2008) Water resources of South Africa, 2005 study (WR2005). WRC Report No. TT381/08. ISBN No. 978-1-77005-813-2. December 2008, Water Research Commission, Pretoria, South Africa.

Midgley, D.C., Pitman, W.V. & Middleton, B.J. (1994) *Surface water resources of South Africa 1990, Volumes I to VI*. Pretoria, South Africa, Water Research Commission. WRC Reports No. 298/1.1/94 to 298/6.1/94.

Pappenberger, F. & Beven, K.J. (2006) Ignorance is bliss: Or seven reasons not to use uncertainty analysis. *Water Resources Research* 42, W0502.

Pitman, W.V. (1973) *A mathematical model for generating monthly river flows from meteorological data in South Africa*. Johannesburg, Hydrological Research Unit, University of the Witwatersrand. Report No. 2/73.

Tshimanga, R.M., Hughes, D.A. & Kapangaziwiri, E. (2011) Initial calibration of a semi-distributed rainfall runoff model for the Congo River basin. *Physics and Chemistry of the Earth* 36, 761–774.

Wagener, T., Lees, M.J. & Wheater, H.S. (2002) A toolkit for the development and application of parsimonious hydrological models. In: Singh, V.P. & Frevert, D. (eds.) *Mathematical models of large watershed hydrology* – Volume 1. Highlands Ranch, CO, Water Resources Publishers. pp. 87–136.

Winter, T.C. (2000) Interaction of groundwater and surface water. In: Duncan, B., Fuentes, R. & Willey, R. (eds.) *Proceedings of the Ground-Water/Surface-Water Interactions Workshop*. Washington, DC. US Environmental Protection Agency, EPA/542/R-00/007. pp. 15–20.

Chapter 10

Groundwater extractions in Flanders – an enforcement review (2005–2009)

J. November, R. Baert & M. Blondeel
Environmental Inspectorate Division; Environment, Nature and Energy Department; Flemish Public Administration, Brussels, Belgium

ABSTRACT

The Environmental Inspectorate Division (EID) is the most important enforcement body for environmental health legislation in the Flemish region (Belgium). Groundwater levels in Flanders are falling in different regions and aquifers. Groundwater is extracted from the different regional aquifers at different depths. The EID is responsible for enforcement of the legislation for the largest (category 1) groundwater extractions as well as all groundwater extractions of companies with the highest environmental impact. Two hundred and ninety groundwater extractions were thoroughly checked in the field for their compliance with the applicable regulations. The percentage of companies with groundwater-related deficiencies dropped from 82% in 2005 to 67% in 2009. At the same time the average number of deficiencies per company increased. Most deficiencies are related to poor construction or maintenance of the production wells as well as poor monitoring of the extraction. The lack of an accreditation process for drilling companies in Flanders causes a lack of assurance that the installation complies with the best available techniques and regulations. Since the available heat in the Flemish underground will be used more and more as a renewable energy source the EID aims for implementation of this accreditation process linked to close on-site monitoring and inspection.

10.1 INTRODUCTION AND STUCTURE OF THE EID

Flanders is situated in the northern part of Belgium, a federal country since the reforms of 1993. With a total surface area of 13 522 km² and 6.2 million inhabitants (January 2008) the population density is 456 inhabitants per km², making it one of the most densely populated areas within Europe. In April 2006, following the reorganisation of the Flemish Public Administration within the framework of Better Administrative Policy, the EID has been part of the Department of Environment, Nature and Energy. Within the framework of the Flemish Parliament Act on environmental licences and Vlarem I, the environmental inspectors of the EID exercise supervision of category 1 establishments. This competence is extended to other supervisory competences resulting from related environmental health legislation, such as the supervision of waste streams.

The EID is constantly striving to improve the quality of enforcement. In this regard, particular attention is devoted to efficient, expert, uniform, integrated and steering action throughout Flanders, and the EID seeks to serve as an example for local authorities.

Besides the EID, many other actors are involved in the enforcement of environmental health legislation, such as mayors, police departments and judicial authorities.

If the enforcement process as a whole is to be successful, these bodies must work together in a constructive manner. The EID is active in establishing networks between these bodies. Another task of the EID is to forge international contacts and to participate actively in international innovations and trends.

For the benefit of policy preparation and policy evaluation, the EID is responsible for advising the Flemish Minister for the Environment on the feasibility and enforceability of the regulations. To this end the EID feeds back its experiences in the field to the policy makers.

Finally, the EID has the task of publicising and providing information about its activities and approach at regular intervals. Through this transparency the EID seeks to create and maintain sufficiently broad-based social support for enforcement.

Within the EID, the Chief Inspectorate fulfils a steering and supporting task and monitors the planning, the depth, the uniform implementation, the harmonisation and the integration of the enforcement campaigns. The Chief Inspectorate is also responsible for the preparation, formulation and evaluation of the enforcement policy. The five local services are mainly responsible for carrying out inspections within their province, taking measures and keeping the company files up-to-date and for coordinating a number of activities.

There is also a horizontal structure which consists of working groups for each environmental compartment. The activities of the working group must guarantee a coordinated and uniform approach throughout the Flemish Region. There are seven active working groups: Waste (including a Waste Chain Team, in charge of monitoring waste collection and transport), Soil and Groundwater, Noise and Vibrations, Genetically Modified Organisms (GMOs), Air, Ozone Depleting Substances and Fluorinated Greenhouse Gases, and Water (Baert, 2008).

10.2 WORKING GROUP 'SOIL AND GROUNDWATER'

The working group 'Soil and Groundwater', within the EID, coordinates all activities concerning inspections of groundwater extractions and the abatement of soil and groundwater pollution. It is composed of a representative from each of the five local services and one representative from the Chief Inspectorate, who acts as a pace-setter.

A minimum of five meetings is held each year by the working group during which inspections are planned and evaluated. Possible bottlenecks, questions from inspectors outside of the working group and legislation updates are typical topics that are discussed. Regular feedback with other authorities by the pace-setter on behalf of the working group is therefore carried out. The findings of the working group are noted in the working group reports, which in turn can be consulted by all members of the EID as a back-up to the oral reports of each of the working group members at their own local service meetings.

Following a period of inactivity, the working group was reactivated at the end of 2004. From that point on it has played a regular part, through different actions and projects, in the annual Environmental Inspection Plan of the EID.

During the years 2005–2009, as well as today, inspections of groundwater extractions constitute the bulk of the activities of this working group. These inspections can be carried out by all inspectors of the EID in the field at the company sites as well as in the office (inspection of the self-monitoring). Therefore specific internal training sessions for all inspectors of the EID on the topic of inspecting drilling and groundwater extraction sites were organised by the working group.

Other past and current activities of the working group, which will not be discussed further in this paper, include creating a methodology for determining soil pollution, inspection of the groundwater-monitoring networks at dump sites, soil and groundwater pollution at shooting ranges and manure storages. The results of all activities of the EID of each past year are published in an environmental enforcement report which is available to the general public.

10.3   GROUNDWATER IN FLANDERS

Groundwater is one of the most valuable natural resources in the Flemish subsoil. Partly due to the high population density mentioned above, demands on this resource are high. Depending on the method, it appears that on average in Flanders and Brussels there is between 1 100 m³ and 1 700 m³ of water (groundwater and surface water) available per person per year. Internationally this is classified as 'very little'. Only a few western countries have even less water per inhabitant (Italy and the Czech Republic). Even in countries such as Spain, Portugal and Greece the water availability per inhabitant is greater than in Flanders and Brussels (Van Steertegem, 2010).

Groundwater is extracted from different aquifers at a depth ranging from several metres up to several hundred metres. Following the stratigraphic time scale, this part of the Flemish subsoil is built up by lower Palaeozoic (Cambrian-Silurian), upper Mesozoic (Cretaceous) and Cainozoic rocks or sediments. On top of the Palaeozoic rock massif, younger sediments have a northerly inclination and mainly consist of chalk, marl, clay or sand. At different depths gravel and sandstone intervals can occur. This post-Palaeozoic stratigraphy of Flanders was created by different ocean-transgression phases throughout the geological time scale. Hence different types of sediments (deep-shallow marine, alluvial, etc.) can nowadays be found.

Dependent on the region in Flanders, the amount of available and usable groundwater varies, for instance, in the western part of Flanders the supplies are very limited. Underneath the thin Quaternary alluvial deposits thick deep-marine clay formations (+150 m) occur. The main source of groundwater in that region lies within Cambrian rock at a depth of several hundred metres. As the demand for groundwater exceeds the availability, this confined aquifer has been overexploited and has lost its artesian nature since the 1970s. Different depression cones have been monitored, desiccation occurs and the water quality is deteriorating. Recent groundwater models predict that drastic measures are required to save this aquifer for the generations to come. On the other hand the Campine basin in the NE part of Flanders contains the major aquifer system of Flanders. Underneath the Quaternary deposits thick (locally +200 m) Neogene, mainly sandy, formations occur. Groundwater in this region is more 'abundant' and a lot easier to extract.

Long-term trends (1999–2009) in groundwater levels throughout the different aquifers were analysed quantitatively and are shown in Figure 10.1 (Van Steertegem, 2010). Approximately 40% showed no statistically significant trend, but there are more decreases of groundwater levels (41%) than increases (19%). Declining groundwater levels therefore

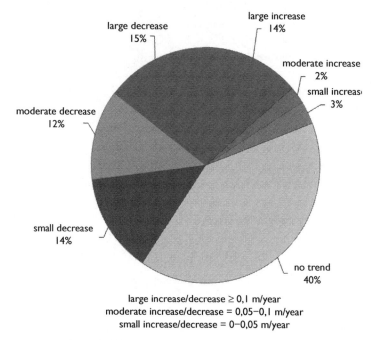

large increase/decrease ≥ 0,1 m/year
moderate increase/decrease = 0,05–0,1 m/year
small increase/decrease = 0–0,05 m/year

Figure 10.1 Long-term trends in groundwater levels.

remain a major problem. In general: the deeper the groundwater level, the more significant the trends – the more rising levels and the larger the declines reported. This is explained by their high reactivity to large groundwater extractions (e.g., depression cones).

Many companies depend on the shallow and deep groundwater supplies on a daily basis via their production wells. Nowadays the number of bores made for purposes other than groundwater extraction (e.g., geothermal bores for heating/cooling of buildings, reconnaissance bores, monitoring wells, etc.) is sharply increasing. Although the use of the groundwater is not the primary purpose of these bores, they often transect the different aquifers and aquitards and can therefore have a big impact on the groundwater systems. Hence the necessity to adequately protect this natural resource.

10.4 GROUNDWATER EXTRACTION

According to Flemish legislation both construction and operation of groundwater bores have to be reported or require an environmental permit. The only exceptions to this rule are the manually operated pumps and groundwater extractions for household purposes not exceeding 500 m³/yr. Depending on the extracted volume, location or origin of the groundwater extraction installations, they are classified as a category 1, 2 or 3 activity. Only category 1 and 2 activities require a direct permit. Category 3 activities only have to be reported to the local authorities. Notwithstanding their classification, some of the category 3 extractions can still extract a significant amount of groundwater in a short period of time. The most frequent examples thereof are groundwater drainages at building or construction sites. They only need to be reported due to their mostly non-permanent nature. Other activities at the establishments (e.g., waste production, air emissions) are categorised in a similar manner to the same 3 categories. The main category of the establishment itself is determined by the highest category among its activities.

For the remediation of contaminated sites, large volumes of groundwater are also extracted annually (e.g., pump-and-treat remediation), but they are permitted via separate specific legislation and do not fall under the jurisdiction of the EID.

At the end of 2009, about 23 000 permitted groundwater extraction installations were reported by the regional government. About 5 000 of these extractions are part of a category 1 establishment, and of these 726 groundwater extraction installations have a permit exceeding 30 000 m³/yr, which is the main criterion that makes most extractions themselves a category 1 activity. Groundwater is extracted from the different regional aquifer systems in the Flemish subsoil, with a depth of the bores ranging from a few metres up to several hundred metres. In general, during the permitting stage, deep groundwater of a very high quality is reserved for high-end purposes (e.g., public water supply, production of foodstuffs, etc.). Groundwater recycling and water-saving measures are promoted. A subsidised 'grey water' scheme has been installed to decrease extraction of the most threatened aquifers.

On a yearly basis about 440×10^6 m³ is permitted for extraction at category 1 or 2 establishments. Public water supply uses the bulk of this volume (60%), followed by agriculture, commerce and other services (Anonymous, 2006). The annual volume effectively extracted from the subsoil is not exactly known, but is estimated at 60% of the permitted volume, which still amounts to 260×10^6 m³. This specifically excludes the volumes extracted for households, soil remediation and the category 3 establishments, since these specific volumes neither have to be permitted nor reported.

The groundwater-permitting system and the constraints therein (annual volume, specific requirements, etc.) are important first steps of groundwater management. They are focused on maintaining and protecting the Flemish groundwater supplies and adjusted as much as possible to the local groundwater situation.

10.5   NECESSITY, PLANNING AND IMPLEMENTATION OF THE INSPECTIONS

The impact on groundwater quantity naturally is related to the extraction volume. Bigger extractions over a long period of time (often at category 1 establishments) therefore have the largest impact. The same analogy cannot be made for groundwater quality. Groundwater extractions also have a large potential impact on the groundwater quality. A major drop of the groundwater levels in semi-permeable aquifers due to local pumping can cause pressure loss and deterioration of the groundwater quality via aeration. This does not necessarily require large volumes to be extracted. Construction errors while installing groundwater bores as well as poor maintenance or decommissioning of the bores can provide perfect 'highways' for pollutants to easily reach the deeper aquifers. A poorly maintained or decommissioned bore often poses a serious risk for the groundwater quality, whatever the volume that is or was extracted.

The acquirement of a groundwater permit, which is usually granted for a period of 20 years, does not necessarily mean compliance with environmental law in the field as will be shown below. The Flemish groundwater supplies (quantity as well as quality) have to be safeguarded now and for future generations. Therefore effective and efficient enforcement of the legislation in the field is an essential part and the cornerstone of groundwater policy in Flanders. 'Those who can richly enjoy the use of groundwater, often don't properly realise its value and are sometimes tempted to use it in an irresponsible manner' (Gullentops and Wouters, 1995).

Within the 5-year time span (2005–2009) 290 groundwater extraction installations were thoroughly checked in the field by the EID for their compliance with the applicable regulations. The exact number of inspections is given in Table 10.1. Starting in 2008, 50 additional inspections were conducted from the office. Companies were asked to send in their self-monitoring data, which were then reviewed by inspectors. If these data were found to be insufficient, field inspections could be planned.

Due to its specific authority, these EID inspections were mainly focused on groundwater extractions at category 1 establishments. The evaluation below therefore does not necessarily apply to the 'smaller' category 2 or category 3 establishments of whom the supervision is the responsibility of the local supervisors at the municipalities. As supervision of all groundwater extraction installations at soil-remediation sites is the sole responsibility of the Public Waste Agency of Flanders, none were included in this selection.

The 290 inspected groundwater extraction installations mentioned above represent about 6% of the 5 000 category 1 establishments where an extraction is present and 40% of the total amount of category 1 extractions (>30 000 m³/yr). For an optimal selection of companies, different criteria, other than its category, are used such as:

- companies or industrial sectors with a poor reputation concerning groundwater extraction;
- companies where initial administrative inspection of the self-monitoring has shown serious flaws;
- recently (re)permitted groundwater extractions;
- illegal or non-permitted groundwater extractions (e.g., via complaints or reports from other authorities, etc.);
- groundwater extractions in overexploited aquifers;
- groundwater extractions pumping more water than permitted;
- groundwater extractions where the permit states special requirements (e.g., monitoring, groundwater research, etc.).

Table 10.1 Inspections of groundwater extraction sites during 2005–2009.

Year	Field inspections	Office inspections
2005	74	–
2006	63	–
2007	56	–
2008	49	25
2009	48	25

In preparation for a field inspection the groundwater permit is reviewed in the office by the inspectors. Special attention is given to all aspects related to groundwater extraction in the permit, e.g., the permitted volume, the permitted depth(s) and aquifer(s), the permitted number of bores, the permitted use of the groundwater and the special requirements. If present in the file, earlier reported data by the company can be reviewed (monitoring data, replacement of a flow meter, etc.) as well as previous inspection reports. Field inspections are sometimes announced beforehand (mostly when inspecting unmanned installations), but more frequently they are not. When arriving on site different aspects in relation to the groundwater extraction are examined via an extensive checklist:

- installed number of bores, their depth and the extracted volume in relation to the permit;
- presence of technical data about the bores and flow meters;
- presence of registers (groundwater flow, groundwater levels and groundwater analysis);
- construction and maintenance of production wells and monitoring wells;
- correct (re)drilling or decommissioning of bores;
- compliance to sector- and special requirements;
- self-monitoring;
- groundwater levy;
- yearly reports to the authorities.

If relevant, water levels are measured by the inspectors and the groundwater is sampled. These samples are sealed and sent to a laboratory for further chemical analysis. During the tour of the company, special attention may be given to other risks of soil or groundwater contamination (e.g., leaking tanks, spillage, etc.). Deficiencies are marked in an inspection report and are remediated via the *modus operandi* available to the EID. In increasing order of importance these are recommendations, exhortations and official reports. When very serious flaws requiring immediate action are found, administrative measures can be taken by the inspectors in the field. These measures may go so far as to seal the installation.

10.6   RESULTS

A bar chart depicting an overview and evolution of the percentage inspected companies with groundwater extraction-related deficiencies is given in Figure 10.2. The last group of bars in this chart shows the percentage of companies where one or more deficiencies were reported. It drops from 82% in 2005 to 67% in 2009. The exact nature and frequency of these deficiencies are shown in the preceding groups of bars. Although these deficiency-percentages are high, it still is a positive trend. It is important to note that not all deficiencies have an equally large environmental impact. Poorly constructed production wells pose a far greater groundwater-pollution risk than the absence of a yearly groundwater analysis report. The presence of illegal groundwater extractions has greater implications for groundwater quantity management than skipping a few of the monthly readings and registering of the flow meters.

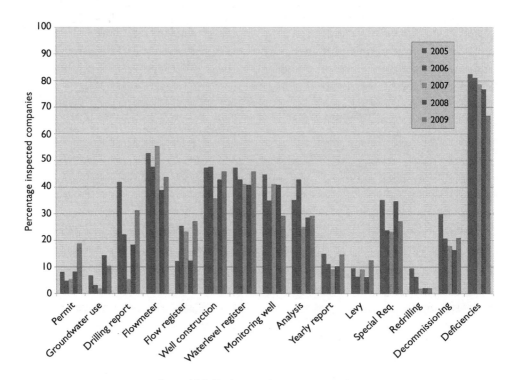

Figure 10.2 Evolution of inspection results.

On the other hand, it can be noted that, in spite of the positive evolution mentioned above, the average number of deficiencies reported per company increased over the same time interval. One explanation of this fact can be found in the increasing experience and expertise of the EID in this matter. Also as a result of these specific inspection campaigns the topic gets a lot more publicity, which in turn can lead to more input about groundwater extractions from third parties (e.g., complaints, reports for other authorities, etc.). With this input, field inspections can be carried out in a more focused, thorough and effective way. This in turn leads to more findings in the field. For instance, in the 2009 inspection campaign special attention was given to illegal, non-permitted, groundwater extraction. This is clearly visible in the first group of bars in Figure 10.2 as a steep incline in the topics 'Permit', 'Drilling report' and 'Flow register'.

Another additional explanation for the high percentages is inherent in the activity itself. Although the importance of the aspect 'groundwater extraction', supported by a focused permitting system, is increasingly emphasised it is just a sideline for most companies. Important exceptions to this rule are, for instance, the public water supply and beverage sectors.

The aspect of extracting groundwater often is, quite literally, one of the most invisible aspects of business. As a result the necessary attention for the extraction process can 'water-down' quickly once the necessary permit is obtained. During inspections operators often have a very hard time to correctly identify and locate production wells or monitoring wells. More so if one or more of the wells has a respectable age or has been decommissioned somewhere in the past. A decommissioned well more often than not rapidly becomes a forgotten well, with all the associated risks. It gets even more complex for operators once different aquifers, pipeline networks and flow meters come into play. Sector requirements of the environmental legislation, or even the specific requirements as stated in the permit, are insufficiently known. Necessary technical or analytical documentation (e.g., calibration certificate of the flow meter, drilling report, chemical analysis report, etc.) often is not at hand and has to be requested from third parties, if at all still available. Changes in company management or personnel often are a source of data-loss about the company's groundwater extraction history. This knowhow should, however, be better maintained and transferred in a company.

The pie chart in Figure 10.3 shows the most common elements included in the written recommendations, exhortations or official reports by inspectors in 2005–2009. Minor deficiencies, remediated by other oral or written communication, are not included in this figure. In relation to both groundwater quantity and groundwater quality different deficiencies can be distinguished.

Concerning groundwater quantity, deficiencies were reported related to monitoring of flow or groundwater levels. Twenty-eight companies extracted significantly larger volumes than was allowed in the permit. At 39 companies a possible bypass of the flow meter was detected. The flow meter itself was insufficiently (re)calibrated at 130 companies. Removal or replacement of flow meters has to be reported immediately to the EID, but 18 companies failed to comply. An equal amount did not monitor groundwater flow as stated in the permit, 105 companies did not have the mandatory amount of monitoring wells and 86 companies insufficiently monitored

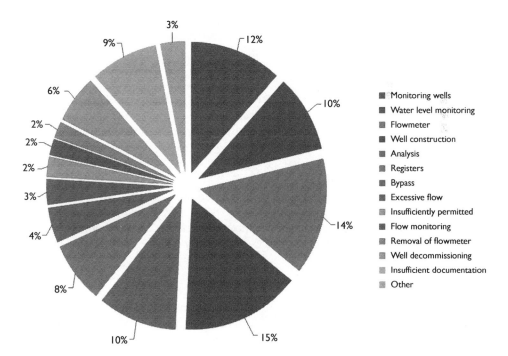

Figure 10.3 Violations reported in 2005–2009.

the groundwater levels. At 61 companies the flow registers and groundwater-level registers were insufficient or absent.

In relation to groundwater quality it is also possible to delineate different deficiencies. Production-well construction at ground level was insufficient at 135 companies, creating a possible risk of groundwater pollution. The mandatory yearly chemical analysis report was missing or incomplete at 90 companies. Old abandoned wells, dangerous potential sources for pollution, were incorrectly decommissioned at 53 companies.

On a more general note, groundwater extraction was not adequately permitted at 22 companies. Groundwater research, explicitly mentioned in the permit, was not conducted or their results were not implemented at 11 companies. Finally, 77 companies were strongly urged to submit the necessary documentation, which was not at hand during the inspection. During the 5 years of these planned inspections 17 companies received a recommendation, 210 companies were strongly urged to remediate the deficiencies and 37 official reports were written (Baert, 2010).

In general, and as depicted by Figures 10.2 and 10.3 most deficiencies can be classified into two main categories. On the one hand deficiencies are mainly related to poor construction or maintenance of the production or monitoring wells. On the other hand the inadequate monitoring of the groundwater extraction (flow, water level, and water quality) is responsible for a large number of deficiencies.

When water is delivered by a public water company, the latter has the responsibility to deliver a constant flow and quality. When companies themselves take care of their own water supply, they should be more aware that they themselves have the same responsibility in terms of flow and quality, be it on a smaller scale. It is exactly that kind of awareness which the EID transfers during these inspections. Good maintenance of the wells, professional monitoring of groundwater flow, level and quality are essential instruments. When carried out correctly the positive trend observed in these inspections will be continued in the future.

10.7 BOTTLENECKS AND RECOMMENDATIONS

As mentioned above the groundwater extracting process is mainly invisible at most companies. Starting at the production well, only the wellhead construction and the initial part of the pipeline network can be visually checked. The main and most important part of the production well is installed in the subsoil up to several hundred metres deep. Once the production well has been constructed, it is virtually impossible for the EID to check whether the correct installation procedure was followed (correct drilling, casement and filter placement, grouting, etc.). Whenever essential documentation (drilling report, construction report, etc.) is missing it is no easy task to verify whether or not the field situation complies with the written facts in the permit or file. The presence of abandoned wells is difficult to trace without this documentation and they are therefore often found by chance while inspecting. Even a 'simple' measurement of the depth of a production well can prove hazardous as the submergible pump and cabling hamper such measurement. In some cases analysis of a groundwater sample can be used to check whether the correct aquifer is being pumped, but often this technique leaves room for speculation and interpretation. Correct placement of the well casing and grouting, installed to prevent short-circuiting between different aquifers, is virtually impossible to check afterwards. Only when inspecting during the installation or decommissioning of wells, structural shortcomings can be more easily detected. Due to the amount of groundwater bores in Flanders such a process would be very technical, expensive, time-consuming and difficult to organise.

Since March 2009 Flemish environmental legislation included a best practices document for drilling, operating and decommissioning of bores for groundwater extraction. Correct procedures could henceforth be enforced by the EID. Four inspections at drilling sites were carried out as a sample in 2009. In three out of four cases the best practices were not followed up correctly. Either the filling of the bores after placement of the casing (with sand, clay or grouting) was not carried out in a proper fashion or sampling and reporting in the field were incorrect. Although legislation now provides the operators with better guidance and guarantees with this 'best practices' document, they themselves still cannot be absolutely certain that the legislation shall be adhered to by the third-party drilling companies; a fact clearly shown by results of these first inspections.

The adherence to environmental law is the responsibility of the operator who holds the permit. In this case the authorities should be able to take direct action against rogue drilling companies, which nowadays is impossible. Since there is a lack of an accreditation process for drilling companies in Flanders, neither their clients nor the government have the assurance that the installation complies with the best available techniques and regulations. Since the available heat in the Flemish underground will be used more and more as a renewable energy source (shallow and deep geothermal drilling, CCS, etc.) the EID aims for an implementation of this accreditation process linked to a close monitoring and inspection in the field. Only accredited drilling companies should be allowed to install any kind of bore. All bores should also have to be reported by the drilling companies to the authorities in order to discourage illegal drilling, which is one of the most important bottlenecks of groundwater management as a whole.

Translating this vision into reality in the future will require input from legislators, groundwater scientists and experts in the technical field. Initial steps in this direction, based on EID publications and input, will be taken in the near future. The EID strongly believes that this is the only way to reduce the (visible and invisible) deficiencies in the field and to preserve the (deep) subsurface and groundwater supplies for generations to come.

REFERENCES

Anonymous (2006) *Grondwaterbeheer in Vlaanderen: het onzichtbare water doorgrond* [*Groundwater in Flanders: Understanding the Invisible Groundwater*]. Aalst, Belgium, Vlaamse Milieumaatschappij.

Baert, R. (2008) *Environmental enforcement report 2007*. Flemish Public Administration, Environment Nature and Energy Department, Environmental Inspectorate Division, Brussels, Belgium.

Baert, R. (2010) *Milieuhandhavingsrapport 2009* [*Environmental enforcement report 2009*]. Vlaamse Overheid; Departement Leefmilieu, Natuur en Energie; afdeling Milieu-inspectie, Brussels, Belgium.

Gullentops, F. & Wouters, L. (1995) Delfstoffen in Vlaanderen [Minerals in Flanders]. Vlaamse Overheid Departement. Brussels, Belgium, EWBL.

Van Steertegem, M. (editor-in-chief) (2010) *MIRA indicator report 2010*. Flemish Public Administration, Flemish Environment Agency, Aalst, Belgium.

Chapter 11

The impact of climate transitions on the radionuclide transport through a sedimentary aquifer

Judith Flügge[1], *Madlen Stockmann*[2],
Anke Schneider[1] & *Ulrich Noseck*[1]
[1]*Gesellschaft für Anlagen- und Reaktorsicherheit (GRS) mbH,
Repository Safety Research Division, Braunschweig, Germany*
[2]*Helmholtz-Zentrum Dresden-Rossendorf e.V., Institute of
Resource Ecology, Dresden, Germany*

ABSTRACT

In long-term safety assessments for nuclear waste repositories in deep formations, geological time scales have to be considered. Possible future climate changes are expected to alter the boundary conditions, the flow regime and the geochemical environment in the aquifers. The codes d³f (Distributed Density-Driven Flow) and r³t (Radionuclides, Reaction, Retardation, and Transport) are being developed to simulate contaminant transport in large heterogeneous areas over long periods of time, considering hydrogeochemical interactions and radioactive decay. A new methodology to use temporally and spatially variant sorption coefficients depending on the geochemical environment is being developed by introducing the transport of relevant components in solution and a pre-computed matrix of sorption coefficients with values being dependent on these components. In this respect, possible future climate changes are being investigated for the reference site Gorleben in Northern Germany. A marine transgression will lead to a decrease of the flow velocities and a horizontal salinity-dependent stratification of the groundwater, while permafrost formation in the upper aquifer and an inflow of glacial melt-water into the lower aquifer will lead to low salinities and high flow velocities in unfrozen zones. Transport simulations employing conventional sorption coefficients are the basis for future analyses employing the new methodology.

Keywords: climate transitions, nuclear waste disposal, radionuclide transport modelling, smart K_d-concept, Northern Germany

11.1 INTRODUCTION AND BACKGROUND

In Germany, high-level radioactive waste is to be disposed of in deep geological formations. As part of long-term safety assessments for radioactive waste repositories, which have to cover a timescale of one million years (BMU, 2010), scenarios have to be considered which lead to the mobilization of radionuclides from the waste and to their transport through the repository system. The internationally acknowledged multi barrier concept comprises a combination of different barriers to obstruct radionuclide migration from the repository to the biosphere: The geotechnical barrier (type of waste, packing, backfill), the isolating rock zone (the geological subsystem of the repository in combination with the geotechnical seals) and the geological barrier (adjoining rock and the overburden), which are expected to have isolating properties for different periods of time (AKEnd, 2002).

This work focuses on the sedimentary overburden of the host rock formation as part of the multi-barrier concept. Many radionuclides are subject to sorption on mineral phases resulting in a deceleration of their transport. Besides the dispersion and dilution, the increased travel time due to retardation leads to a reduction of their concentration in near-surface aquifers and thus of the radiation exposure because of additional radioactive decay of many radionuclides during transport. On the other hand, retarding the transport of a mother nuclide could lead to an elevated generation of daughter nuclides and, in case the daughter nuclide is dose relevant, to an elevated radiation exposure. Describing the sorption as realistically as possible is therefore one of the crucial aspects in long-term safety assessments.

Spatially and temporally constant sorption coefficients, K_d-values, for each hydrogeological unit are currently used in transport codes for groundwater modelling. However, temporally and spatially variable geochemical conditions could lead to significant changes in the sorption of different radionuclides. Therefore, a description is required of the sorption as a transient parameter depending on environmental parameters. In order to deal with the variable sorption, an advanced methodology in the form of the newly established 'smart K_d-values' in codes for long-term safety assessments is being developed and tested. Long-term climate changes may lead to, among others, a transgression of the sea and the inundation of the respective site, or to the formation of permafrost in the sedimentary overburden with an inland ice sheet in close vicinity. Geochemical changes are then caused by an infiltration of seawater with a higher salinity than groundwater or of glacial melt-water with a low salinity. Groundwater flow and transport simulations are performed to study the impact of selected climate transitions on a potential repository site for radioactive waste in Northern Germany.

The background, assumptions and realisations of transport simulations employing the smart K_d-concept for the reference site are introduced in sections 1 and 2. The implementation of the smart K_d-concept is still in progress. Therefore, results of the first orienting simulations employing the conventional K_d-values are presented and discussed in section 3 and 4.

11.1.1 The Gorleben Reference Site

Since 1979, the Gorleben site in Lower Saxony, Northern Germany (Figure 11.1), has been investigated by the Federal Institute for Geosciences and Natural Resources (BGR) and several other research organisations to determine its suitability to host a nuclear waste repository (Klinge *et al.*, 2007; Köthe *et al.*, 2007). A comprehensive dataset is available on the geology and hydrogeology of the Gorleben salt dome and its overburden. Therefore, it was selected as a reference site for this work. The Gorleben-Rambow salt structure has a length of ca. 30 km and a width varying between 1.5 km and 4 km. It strikes SW-NE and is crossed at the surface by the Elbe River and its tributaries, the Löcknitz River and the Seege River (Klinge *et al.*, 2007).

The Gorleben Area

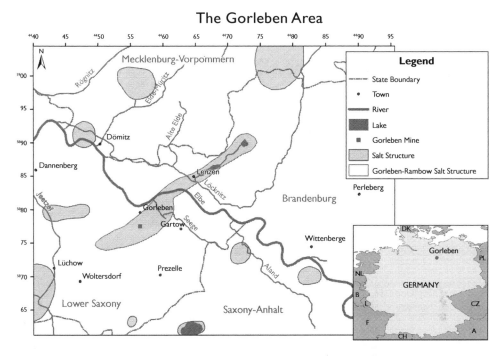

Figure 11.1 The Gorleben area: The Gorleben-Rambow salt dome is located in Northern Germany at the state boundaries of Mecklenburg-Vorpommern, Brandenburg, Saxony-Anhalt and Lower Saxony. It is one of many salt structures in the area. The salt structure has a length of ca. 30 km and a varying width of between 1.5 km and 4 km. It strikes SW-NE and is crossed at the surface by the Elbe River and its tributaries, Löcknitz and Seege Rivers. Abbreviations in lower right overview map: DK: Denmark, PL: Poland, CZ: Czech Republic, A: Austria, CH: Switzerland, F: France, L: Luxembourg, B: Belgium, NL: Netherlands.

11.1.1.1   Geology and hydrogeology

The Tertiary and Quaternary sedimentary overburden of the salt dome and its neighboring rim synclines form a system of aquifers and aquitards (Figure 11.2) up to 430 m thick (Klinge *et al.*, 2002). The basis of the regional flow system is represented by the Tertiary Rupel Clay. The Tertiary Lower Brown Coal Sands and Eochattian silts and the Quaternary Elsterian channel sands form the lower aquifer (Ludwig and Kösters, 2002). The intercalated aquitard is a set of the Tertiary Hamburg Clay and the Quaternary Lauenburg Clay Complex, which is superposed by mostly Weichselian and Saalian sediments representing the upper aquifer (Klinge *et al.*, 2002). Main structural elements of the system are the salt dome itself with the adjacent north-western and south-eastern rim synclines, a contact of the salt dome to the lower aquifer in the glacial melt-water channel, the so-called Gorleben channel, formed by glacial erosion during the Elsterian cold stage, and hydraulic connectivities between the two aquifers, so-called hydraulic windows, likewise formed by glacial erosion (Klinge *et al.*, 2007).

The hydraulic system can be roughly divided into a lower aquifer, an intercalated aquitard and an upper aquifer. The Zechstein salt is in contact with the Quaternary

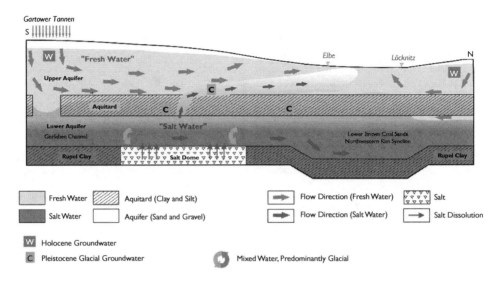

Figure 11.2 Schematic cross-section of the hydrogeological system at the reference site Gorleben, ten times vertical exaggeration (modified from Klinge *et al.*, 2002, with permission from BGR): The system consists of an upper freshwater aquifer, an intercalated aquitard and a lower salt-water aquifer. Salt dissolution at the contact to the salt dome in the lower aquifer causes a density-driven groundwater flow. From the salt dome, part of the salt water rises up to the surface in the area of the Elbe River, part of it sinks into the north-western rim syncline due to its higher density. Groundwater recharge at the surface and an inflow from the north into the lower aquifer are the sources of freshwater in the system. In general, the salt water has a Pleistocene signature, while the freshwater has a Holocene signature.

aquifer in the Gorleben Channel, and salt can therefore dissolve in the groundwater. The groundwater shows stratification with an upper freshwater body and a lower salt-water body (Ludwig and Kösters, 2002). The distinct relief of the basis of the freshwater body is governed by the relief of the underlying aquitard as well as by the regional groundwater flow. Groundwater recharge takes place in the area of the Gartower Tannen in the south and in the lowlands north of the Elbe River. The highest freshwater thicknesses occur in those areas of the rim synclines where the salt-water sinks to the bottom of the system due to its higher density as well as where the aquitard is missing. The lowest thicknesses are found in the entire area of river beds and lowlands north of the Elbe River and above the Gorleben salt dome. The salt-water/freshwater interface rises from the Gorleben Channel to the north and reaches the water table close to the Elbe River (Klinge *et al.*, 2007; Ludwig and Kösters, 2002).

11.1.2   Climate transitions

Possible future climate transitions can be derived from the geological past. The Quaternary period is characterised by extreme climate changes, i.e. the repeated advance and retreat of inland ice sheets, and the transgression of the North Sea in Northern

Germany due to a global sea-level rise (Benda, 1995). With the knowledge about past climate and its driving forces, possible future climate states can be simulated.

Several studies investigate the possibility of a future sea-level rise due to global warming. Ganopolski (Ganopolski, PIK Potsdam, personal communication, 2007) assumes a sea-level rise of 20 m to 30 m within the next 50 000 years. Taking the thermal expansion and density changes of seawater into account, the sea-level rise due to the melting of the global ice volume can be calculated. Therefore, a sea-level rise of up to 80 m is possible (Williams and Hall, 1993), but is not regarded as a realistic scenario (Meyer, TU Braunschweig, personal communication, 2006). According to different publications, the next cold stage may be expected in ca. 20 000 years (Emiliani, 1957), not earlier than 50 000 years (Loutre and Berger, 2000) or even not earlier than 170 000 years to 500 000 years (Archer and Ganopolski, 2005). Values are based on different methods, such as the evaluation of oxygen isotopes of foraminifers from deep-sea cores or modelling approaches considering changes in orbital parameters and different assumptions, e.g. regarding the future CO_2 emissions.

All in all, a repetition of past interglacial and glacial cycles is most probable on a time scale of one million years, so that a future cold stage or a sea-level rise will have to be considered for the long-term future.

11.1.2.1 Seawater transgression

Quaternary sea-level oscillations can be regarded as a result of the Milankovitch cycles and hence are dependent on the global ice volume (Imbrie *et al.*, 1984). In Northern Germany, the sea-level variations amounted to several decametres up to ca. 120 m in the course of the past 18 000 years (Streif, 2004). Several Quaternary North Sea transgressions are recorded for Northern Germany. In the geological past, periods of seawater inundation in Northern Germany persisted for a few thousand years only (Streif, 2004). The maximum extension of the Quaternary marine transgressions occurred during the Holstein Warm Stage, which lasted from 335 000 to 330 000 years Before Present (BP). The marine transgression occurred in the form of a single, uninterrupted sea-level rise with an average rising rate of 1 m per 100 years (Streif, 2004). This led to a local sea-level rise of more than 50 m (Streif, 2004) up to ca. 65 m (Linke *et al.*, 1985), while the global sea-level rise amounted to more than 100 m (Rohling *et al.*, 1998). The sea-level high-stand persisted for ca. 5 000 years.

11.1.2.2 Permafrost

In the geological past, climate states with the formation of permafrost occurred periodically at the Gorleben site. According to model calculations, permafrost during the Weichsel Cold Stage in the Gorleben area reached thicknesses of between 40 m and 140 m (Klinge *et al.*, 2007). Here, periods with and without permafrost alternated with a periodicity of ca. 10 000 to 30 000 years. Completely unfrozen zones – so-called taliks – could have been formed in areas beneath rivers and lakes due to their thermal influence (Delisle, 1998; Keller, 1998). In the case of a future glacial stage it has to be assumed that there will be, similar to that of the last glacial stage (Weichselian), an inland ice sheet north or northeast of the Gorleben area. During the maximum extension of the Weichselian ice sheet ca. 16 000 to 20 000 years ago (Streif, 2004), the ice margin was located in the immediate vicinity of the investigation area, less than 50 km away (Kösters *et al.*, 2000). Permafrost areas melt in the case of glacial over-riding due to a sub-glacial thermal disequilibrium (Boulton and Hartikainen, 2004). Therefore, the permafrost decays below the inland ice sheet. Glacial melt-water is forced into the sub-glacial aquifer, and will rise to the free surface beyond the perma-frost in front of the ice sheet. In this case, large volumes of melt-water could infiltrate into the underground and flow into the lower aquifer of the model domain. According to Boulton (Boulton *et al.*, 1993) unfrozen zones in the permafrost may concentrate discharge and produce exceptionally high upward water fluxes. The Elbe River, which served as a glacial valley with a talik during the Weichselian glacial stage, is expected to persist and, due to the high melt-water volumes from the ice sheet, to have a larger width than today (Kösters *et al.*, 2000). Other taliks can be assumed to have formed beneath the smaller tributaries of the Elbe River. During rainfall or flood events, these taliks serve as infiltration areas for groundwater recharge. At the same time, ground-water could be discharged due to high inflow from the north into the lower aquifer, through the taliks of the upper aquifer, up to the surface.

11.2   METHODS

In order to model the impact of climate transitions on the radionuclide transport, existing codes were further developed and the smart K_d-concept was implemented.

11.2.1   Groundwater flow and transport codes

Flow and transport were simulated using the codes d³f for groundwater flow and r³t for radionuclide transport, which were developed under the auspices of GRS (Fein, 2004; Fein and Schneider, 1999). The main characteristic of d³f is the possibility to calculate the transient, density-driven transport of salt, which is the most important process influencing the groundwater flow in the model area (Klinge *et al.*, 2007). Transport simulations with r³t are based on flow fields calculated by d³f. The most important feature of r³t is the possibility to simulate pollutant transport in very large model domains for long periods in time.

11.2.2   The smart K_d-concept

First of all, a comprehensive dataset about the mineralogical composition of the sedi-mentary overburden and the groundwater chemistry of the reference site had to be compiled. The most important minerals of the three hydrogeological units are quartz, feldspar, muscovite, gibbsite, goethite, calcite, kaolinite, and illite in different compo-sitions of weight (Köthe *et al.*, 2007). Apart from the mineral composition of the sedi-mentary overburden, the sorption behaviour of the radionuclides is influenced by the groundwater composition, since cations might act as concurrent sorbates and anions as ligands thereby affecting the radionuclide concentration in solution. In our system, the cations Ca^{2+}, Na^+, K^+, Fe^{3+} and Al^{3+} and the complexing ligands CO_3^{2-}, SO_4^{2-}, and SiO_3^{2-} are assumed to be the most important components. Redox effects are not taken into account so far. In highly mineralised waters Cl^- becomes also important, e.g. in the lower aquifer. Based on their half-lives, sorption characteristics, and availability of comprehensive and reliable thermodynamic data, the following radionuclides were selected to be included in the calculations: Cs-135, Ni-59, Se-79, Np-237, U-238 (with the daughter nuclides U-234, Th-230, Ra-226) and Cm-247 (with the daughter nuclides Am-243, Pu-239, U-235).

For the investigations and calculations, a dedicated thermodynamic database based on international efforts like the Nagra/PSI Chemical Thermodynamic Data Base Version 01/01 (Nagra/PSI 2002) was created. Thermodynamic sorption data for representative sorbates (pair of element and mineral) are taken from the thermodynamic sorption database RES^{3}T (Rossendorf Expert System for Surface and Sorption Thermodynamics) (Brendler *et al.*, 2003). Where datasets were not available, batch experiments were performed in order to derive the necessary data (mono-, bi-, and trivalent cations onto the minerals orthoclase and muscovite).

In our concept (Noseck *et al.*, 2012), the sorption of radionuclides onto a sediment (a defined mixture of mineral phases) is based on specific environmental parameters such as pH-value, DIC (Dissolved Inorganic Carbon), ionic strength (salt concentration), concentration of calcium, and total radionuclide concentration. Taking these environmental parameters into account, the smart K_d-values are calculated using the geochemical modelling code PHREEQC Version 2.17.01 (Parkhurst and Appelo, 1999) and saved as K_d-matrices, which can be read by the transport code r^{3}t. During the simulation, r^{3}t calculates the environmental parameter values for each time step and point in space. The ionic strength is provided by the flow simulation with d^{3}f. Based on these values, r^{3}t accesses the K_d-matrices, and retrieves the corresponding K_d-value for each point in time and space.

11.2.3   Groundwater flow and transport model

The model is based on a schematic cross-section through the Gorleben area, which summarises all main geological and hydrogeological characteristics of the site under present conditions (Figure 11.2).

11.2.3.1   Model set-up

The model comprises three geological units, representing the lower aquifer, the aquitard, and the upper aquifer. The lower aquifer is characterised by high salt concentrations of the groundwater resulting from the contact to the salt dome, while the upper aquifer is characterized by low salt concentrations. Groundwater recharge takes place at the surface in the area of the Gartower Tannen and north of the Elbe River. Part of this water is drained to the lowlands of the Elbe River, while the other part may infiltrate through the hydraulic windows of the aquitard down to the lower aquifer. As the salt dome is in contact with the lower aquifer, salt is dissolved in the groundwater, with concomitant increases in the salt concentration and the density of the groundwater. From the contact with the salt dome, the higher salinity water flows northwards and sinks in the north-western rim syncline. Apart from that it can rise up into the upper aquifer through hydraulic windows or through the aquitard, resulting in locally elevated layers of salt water underlain by less saline water and salt-water occurrences at the surface.

11.2.3.2 Model geometry

A two-dimensional model was set up based on the schematic cross-section (Figure 11.2) and the geological and hydrogeological information given by numerous publications (e.g. Klinge *et al.*, 2007; Klinge *et al.*, 2002; Ludwig and Kösters, 2002). The two-dimensional model has a length of 16.4 km and a maximum depth of 400 m (Figure 11.3). It consists of three different units representing a lower and an upper aquifer with an intercalated aquitard. In the model, both aquifers have a thickness of 100 m and the north-western rim syncline is simulated with an additional thickness of the lower aquifer of 150 m. The aquitard has a thickness of 50 m. Both the Hamburg Clay and the Lauenburg Clay Complex are locally interrupted by hydraulic windows, i.e. local absence of the aquitard. One hydraulic window with a width of 500 m is simulated close to the southern boundary of the model and a second hydraulic window with the same width at the northern boundary of the north-western rim syncline. The contact of the lower aquifer to the cap rock of the Gorleben salt dome is marked in red in Figure 11.3.

11.2.3.3 Flow model

Values for the hydraulic parameters were taken from different publications about the Gorleben aquifer system (cited in Flügge, 2009). The permeability k is $1 \cdot 10^{-12}$ m² for the aquifers and $1 \cdot 10^{-16}$ m² for the aquitard. The porosity φ is 0.2 (aquifers), and 0.05 (aquitard). A uniform longitudinal dispersion length α_L of 10 m, a transverse dispersion length α_T of 1 m and a molecular diffusion coefficient D_m of $1 \cdot 10^{-9}$ m²/s is set for the entire model domain. The temperature rises linearly from 8°C at the surface to 20°C at the bottom of the model. Heat transport is not calculated in the simulations. Boundary conditions for each climate transition are described in the following sections.

11.2.3.3.1 Seawater transgression

As a basis for further model simulations, the present flow field and salt distribution in the Gorleben area was modelled first. The present conditions have evolved since the

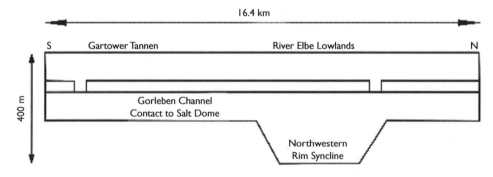

Figure 11.3 Geometry of the groundwater flow and transport model, ten times vertical exaggeration: The model consists of two aquifers and an intercalated aquitard. Main structural elements, such as the north-western rim syncline, the Gorleben channel with the contact to the salt dome, and two hydraulic windows are represented in the model geometry. The model has a length of 16.4 km and a maximum depth of 400 m.

end of the Weichsel Cold Stage and the beginning of the Holocene at 11,500 years BP (Streif, 2007; Klinge *et al.*, 2007). The hydrogeological system is not yet in a steady-state. Initial conditions for the salt concentration are set according to the presumed salt distribution in the Gorleben area at 11 500 years BP. Boundary conditions were defined according to those given in various publications (Klinge *et al.*, 2007; Klinge *et al.*, 2002; Ludwig and Kösters, 2002). At the southern and the northern part of the model surface, groundwater recharge is assumed. In the centre of the surface, a hydrostatic pressure is defined. Here freshwater may infiltrate or groundwater of different salinity may be discharged. At the northern boundary, an inflow of groundwater is defined. After 11 500 years model time, the present climate state is reached. The model was run for another 150 000 years model time. Then the lateral boundaries were closed and a time-dependent pressure boundary condition was defined at the surface representing a transgression for 5 000 years, a sea-level high-stand at 50 m for another 5 000 years, and a regression for another 5 000 years.

11.2.3.3.2   Permafrost

In the model, permafrost regions in the upper aquifer are simulated by reduction of the permeability of $k = 1 \cdot 10^{-12}$ m^2 to $k = 1 \cdot 10^{-20}$ m^2. The permafrost extends over the entire thickness of the upper aquifer of 100 m. A large unfrozen zone with a width of 5 km is located in the area of the Elbe lowlands above the north-western rim syncline. Another smaller unfrozen zone is located below the Seege River at 450 m to 550 m north of the southern boundary. All other hydrogeological parameters remain unchanged from the present climate. At the model surface, a Dirichlet condition is defined for the pressure representing hydrostatic conditions. As specified for the present climate state, freshwater may infiltrate or groundwater of different salinity may be discharged at the surface. These conditions were applied for the southern boundary of the lower aquifer as well. At the northern boundary of the lower aquifer, a maximum inflow of melt-water is assumed which is given in form of a head gradient of 0.1 (Kösters *et al.*, 2000). The simulation starts with present conditions and runs for 150 000 years. Then, a freezing and thawing of permafrost for a time of 40 000 years was simulated by adapting the permeability of the upper aquifer according to calculations of past permafrost growth and decay (Delisle, 1998) and by adapting the boundary conditions to the present ones depending on the respective permafrost thickness. The permafrost state is assumed to be followed by present conditions, so the model run terminates with another 15 000 years of present climate conditions.

11.2.3.4   Transport model

In the case of an altered evolution of the repository, radionuclides can be released from the near field into the far field, be transported through the overburden, and reach the biosphere. The radionuclides are released into the model area at the centre of the contact to the salt dome in form of a stylised inflow function derived from previously conducted model simulations for the Gorleben site (Keesmann *et al.*, 2005). Since this is a case study, it is not the aim to model realistic scenarios but instead to identify the impact of altering geochemical conditions in the hydrogeological system due to climate changes. In order to observe these impacts in the entire model domain, a maximum distribution of the radionuclides is aimed at by assuming the inflow of radionuclides to start with the first time step. The inflow of radionuclides is given at the centre of the salt dome. For the other boundaries, inflowing water always has a radionuclide concentration of $c_{rn} = 0$ mol/m³, while for an outflow, c_{rn} is set equal to the concentrations in the outflowing groundwater.

For transport calculations, the mineral composition of the three hydrogeological units had to be quantified. Different fractions of the mentioned minerals (see section 11.2.2) were set for the three hydrogeological units according to Köthe (Köthe *et al.*, 2007), e.g. the lower aquifer doesn't show any calcite content, while the aquitard is the only layer with an illite content.

Additionally, boundary conditions had to be defined for the pH value, the DIC and the Ca^{2+} concentration. For the present state a Dirichlet condition is set for the inflow into the lower aquifer. The inflowing groundwater is assumed to have the same chemical composition as the formation water of the lower aquifer (Klinge *et al.*, 2007). At the model surface the chemical composition of the water depends on the flow direction, which was determined in the d³f calculations. In case of an outflow, it is given by the chemical composition of the formation water. In case of an inflow, the chemical composition is defined by the composition of the precipitation (Mattheß, 1994).

11.2.3.4.1   Seawater transgression

For the marine transgression, the chemical composition of the surface water is dependent on the flow direction. In case of an outflow, the chemical composition is given by the formation water. In case of an inflow the seawater composition is set according to Langer (Langer and Schütte, 2002). Stepwise functions are defined for the surface boundary condition in the transition states between the present climate and the seawater transgression. For the transitions, the lateral boundaries of the model are defined as impermeable.

11.2.3.4.2   Permafrost

For the permafrost state, the boundary condition for the present climate state at the model surface applies. At the southern boundary of the lower aquifer outflowing groundwater has the chemical composition of formation water. The chemical composition of inflowing glacial water from the north into the lower aquifer is derived from the chemical composition of an Antarctic ice core (Legrand *et al.*, 1988). For the transition between permafrost and the present climate the boundary conditions are interpolated using stepwise functions or defined based on the direction of the groundwater flow, which was calculated by d³f.

11.3   RESULTS AND DISCUSSION

The development of the smart K_d-concept and its implementation in the codes d³f and r³t are still in progress. In a previous project, flow and transport simulations using the conventional K_d-concept were performed for different climate states including the permafrost state and the seawater inundation (Flügge, 2009; Noseck *et al.*, 2009). These calculations were the basis for the time- and geochemistry-dependent simulations to be conducted within this project. Climate transitions were not regarded; the boundary conditions for the flow model were adjusted to the different climate states in one single time step. Transport simulations were run on constant flow fields over a period of one million years of model time, as the coupling of the flow and the transport code were not yet available. Amongst other simulations, an inflow of one mol per radionuclide in the form of a delta pulse was assumed. Conventional K_d-values were taken from a data set with average salinity-dependent values for argillaceous and sandy material (Suter *et al.*, 1998), which were gained from batch experiments using Gorleben sediment samples.

11.3.1 Flow simulations

After modelling the present climate state, a seawater inundation was simulated until steady-state was reached (Figure 11.4, salt concentration is given as relative salt concentration c_{rel}). The simulation resulted in a horizontal layering of the groundwater. The advective groundwater flow ceases, and convection processes cause a mixing of waters with varying salinity. Since there is no inflow into the lower aquifer, the flow is driven by mixing processes between the groundwater with a higher salt concentration above the salt dome contact and the groundwater with lower salt concentrations in the central and northern area of the lower aquifer. An inundation time of only a few thousand years has a clear influence on the salt concentrations in the upper aquifer and in the aquitard which approach the concentration of seawater. During this period, the salt concentration of the lower aquifer is hardly affected by the changed boundary conditions.

Main features of the permafrost climate state (Figure 11.5) are the high velocities in the lower aquifer, the slow transport processes by diffusion in the frozen zones of

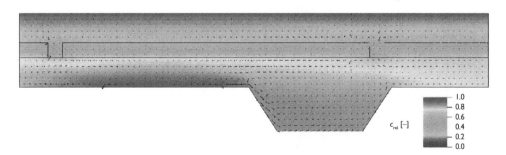

Figure 11.4 Relative salt concentration c_{rel} (colours) and velocity field (vectors) for the seawater inundation after 600 000 years of modelling time, ten times vertical exaggeration: The groundwater shows a horizontal stratification depending on the salinity and density of the groundwater. The advective groundwater flow ceases, and convection is the driving force of the groundwater flow. A convection cell can be observed between the contact to the salt dome in the lower aquifer and the north-western rim syncline. Colours indicate the relative salt concentration (c_{rel} = 1 represents saturated brine, c_{rel} = 0 represents freshwater), and the velocity field is given by the vectors. The length of the velocity vectors is proportional to the velocity; vectors are cut off at 0.036 m/a.

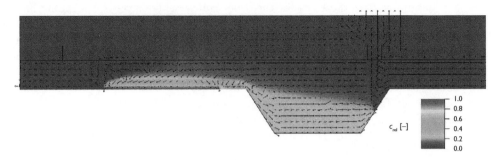

Figure 11.5 Relative salt concentration c_{rel} (colours) and velocity field (vectors) for the permafrost after 600,000 years model time, ten times vertical exaggeration: Highest flow velocities can be observed in the lower aquifer due to the large inflow of glacial melt-water into the lower aquifer. In contrast to that, the advective flow vanishes in the frozen areas of the upper aquifer, where the salt transport is dominated by diffusion. The salt concentration of the groundwater is strongly reduced in the entire model domain. The groundwater flow is directed from the north of the lower aquifer through the hydraulic window to the surface as well as through the lower aquifer to the south. Colours indicate the relative salt concentration (c_{rel} = 1 represents saturated brine, c_{rel} = 0 represents freshwater), and the velocity field is given by the vectors. The length of the velocity vectors is proportional to the velocity; vectors are cut off at 0.036 m/a.

the upper aquifer and the strongly reduced salt concentrations in the entire model domain, in particular in the north-western rim syncline. The groundwater flow in the lower aquifer is directed to the south, while there is a residual salt concentration remaining in the north-western rim syncline.

11.3.2 Transport simulations

Results for the transport of Cs-135 are shown exemplarily for the two climate states. For a seawater inundation (Figure 11.6), the flow velocity is considerably reduced and diffusion dominates both the mixing processes between freshwater and saline water and the migration of radionuclides. After a modelling time of 1 000 years, Cs-135 reaches the basis of the aquitard. After one million years, Cs-135 can be observed at the surface. It is remarkable that the transport of Cs-135 by diffusion proceeds almost isotropically due to the low groundwater flow velocities. Cs-135 leaves the model area approximately vertically above the salt dome contact at the surface.

For the permafrost climate state (Figure 11.7), high flow velocities in the unfrozen zones and transport by diffusion in the frozen zones can be observed. In the permafrost regions within the upper aquifer, the transport by advection is slowed down drastically. Moreover, due to the increased sorption in the aquitard, the upward transport is strongly retarded. Cs-135 is not transported into the north-western rim syncline and does not reach the model surface after one million years.

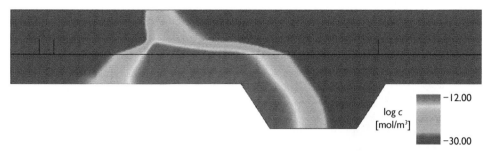

Figure 11.6 Concentration distribution of Cs-135 in the model area after one million years of modelling time for the seawater inundation, ten times vertical exaggeration: Diffusion is the dominant transport process, Cs-135 shows a radial distribution from the contact to the salt dome, where the inflow into the model domain is located. Highest Cs-135 concentrations at the model surface are observed approximately vertically above the salt dome. Colours indicate the Cs-135 concentration on a logarithmic scale.

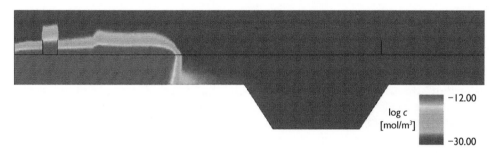

Figure 11.7 Concentration distribution of Cs-135 in the model area after one million years of modelling time for the permafrost climate state, ten times vertical exaggeration: According to the groundwater flow direction in the lower aquifer, Cs-135 is transported to the south and leaves the model domain still within the lower aquifer. In the permafrost regions within the upper aquifer, the transport by advection is slowed down drastically. Here, transport by diffusion dominates. Due to lower salinities, the sorption capacity increases and the transport of Cs-135 is retarded. It is not transported into the north-western rim syncline and does not reach the model surface after one million years. Colours indicate the Cs-135 concentration on a logarithmic scale.

Generally, the climate states determine the groundwater flow field and thus the flow direction and the dominant transport processes (advection or diffusion), respectively. Accordingly, the results for the radionuclide transport show considerable differences. The different climate states have a low influence on the strongly sorbing radionuclides, e.g. Zr-93, while for the weaker sorbing radionuclides the retardation is strongly influenced by the flow direction and velocity, the transport process and the mineralization of the groundwater. Introducing a temporally and spatially variable K_d-value, which takes the hydrogeochemistry into account, is a crucial step towards a more realistic treatment of the sorption in long-term safety assessments.

11.4 CONCLUSIONS AND OUTLOOK

In long-term safety assessments, a time frame of one million years has to be considered. Climate changes have to be regarded for long-term evolution of a repository site. Long-term climate changes may lead to, for example, the inundation of the respective site, or to the formation of permafrost. Aiming at developing and testing a new methodology to include temporally and spatially variable sorption coefficients in codes for long-term safety assessments, the smart K_d-concept is introduced, which allows direct determination of the impact of changing geochemical conditions on the sorption coefficient. Therefore, the transport of components affecting the K_d values in solution will be regarded in transport calculations. Depending on environmental parameters, the transport code r³t accesses a pre-computed K_d-matrix, where appropriate values are stored.

Model calculations employing the smart K_d-concept are currently conducted for the reference site Gorleben. The hydrogeological system is well known and model assumptions were tested and reviewed in former projects. Based on general assumptions about the driving forces of climate and future climate evolution, two climate transitions were chosen to be investigated within this work: A marine transgression and permafrost formation.

Based on past investigations presented in this paper, the hydraulic and geochemical impact of the climate transitions can be estimated. During a marine transgression the groundwater flow ceases and diffusion controls the distribution of contaminants. Accordingly, the radionuclide transport is slow. Radionuclides are transported by radial diffusion (and advection) from the source and reach the model surface vertically above the salt dome. The periglacial permafrost state is caused by an inland ice sheet which is in close proximity during a cold stage. High flow velocities in the lower aquifer are observed for this climate state, where advection is the dominant transport process. Radionuclides are transported out of the model domain in the lower aquifer. Diffusion dominates the transport through the frozen upper aquifer. Freshwater presence in parts of the model domain induces a higher sorption of the radionuclides.

Model calculations using the new methodology will improve the quantification of the impact of climatic transitions on the radionuclide transport in the sedimentary overburden of salt domes. In previous model calculations, transport simulations were conducted on constant flow fields. Taking the temporal and spatial geochemical variations into account, it is crucial to base the transport calculations on transient flow fields. The results of the model calculations using the smart K_d-concept will be compared to those employing the conventional K_d-values. The spatial distribution of the considered radionuclides, the maximum radionuclide concentrations at the surface and their location will be evaluated. Based on the simulation result, the smart K_d-concept will be extended in the future in order to consider other important components.

ACKNOWLEDGEMENTS

This work was financed by the German Federal Ministry of Economics and Technology (BMWi) under contract numbers 02 E 10518 and 02 E 10528. The authors would like to thank Vinzenz Brendler (Institute of Resource Ecology, Helmholtz-Zentrum Dresden-Rossendorf) for fruitful discussions and Michael Lampe (Goethe Center for Scientific Computing (G-CSC), Goethe University Frankfurt) for his work in implementing the smart K_d-concept in the transport code r³t. Thanks also to the two anonymous reviewers for their valuable input.

REFERENCES

AkEnd (2002) *Auswahlverfahren für Endlagerstandorte. Empfehlungen des AKEnd – Arbeitskreises Auswahlverfahren Endlagerstandorte (Site selection procedure for repository sites. Recommendations of the AKEnd – working group selection procedure for repository sites).* Final report, Germany.

Archer, D. & Ganopolski, A. (2005) A movable trigger: Fossil fuel CO_2 and the onset of the next glaciation. *Geochemistry Geophysics Geosystems*, doi: 10.1029/2004GC000891.

Benda, L. (1995) *Das Quartär Deutschlands (The Quaternary in Germany).* Berlin, Germany, Gebrüder Bornträger.

BMU (2010) Sicherheitsanforderungen an die Endlagerung wärmeentwickelnder radioaktiver Abfälle (Safety requirements for the final disposal of radioactive waste). Bundesministerium für Umwelt, Naturschutz und Reaktorsicherheit (German Federal Ministry for the Environment, Nature Conservation and Nuclear Safety), Berlin, Germany.

Boulton, G.S. & Hartikainen, J. (2004) Thermo-hydro-mechanical impacts of coupling between glaciers and permafrost. In: Stephansson, O. Hudson, J.A. & Jing, L. (eds.) *Coupled therm o-hydro-mechanical-chemical processes in geo-systems – fundamentals, modelling, experiments and applications*, Volume 2. Amsterdam, The Netherlands, Elsevier. pp. 293–298.

Boulton, G.S., Slot, T., Blessing, K., *et al.* (1993) Deep circulation of groundwater in overpressured subglacial aquifers and its geological consequences. *Quaternary Science Reviews* 12, 739–745.

Brendler, V., Vahle, A., Arnold, T., Bernhard, G. & Fanghänel, T. (2003) RES³T-Rossendorf expert system for surface and sorption thermodynamics. *Journal of Contaminant Hydrology* 61, 281–291.

Delisle, G. (1998) Numerical simulation of permafrost growth and decay. *Journal of Quaternary Science* 13, 325–333.

Emiliani, C. (1957) Temperature and age analysis of deep-sea cores. *Science*, 125, 383–387.

Fein, E. (2004) *Software package r³t – Model for transport and retention in porous media.* Braunschweig, Germany, GRS. Final Report. BMWA-FKZ 02E9148/2.

Fein, E. & Schneider, A. (1999) *d³f – Ein Programmpaket zur Modellierung von Dichteströmungen (Software package modeling density-driven flow).* Braunschweig, Germany, GRS. Final Report. BMWi-FKZ 02C04650.

Flügge, J. (2009) *Radionuclide Transport in the Overburden of a Salt Dome – The Impact of Extreme Climate States.* München, Germany, Verlag Dr. Hut.

Ganopolski. PIK Potsdam (Personal communication, 2007).

Imbrie, J., Hays, J.D., Martinson, D.G., McIntyre, A., Mix, A.C., Morley, J.J., Pisias, N.G., Prell, W.L. & Shackleton, N.J. (1984) The orbital theory of pleistocene climate: Support from a revised chronology of the marine $\delta^{18}O$ record. In: Berger, A.L., Imbrie, J., Hays, J., *et al.* (eds.) *Milankovitch and climate.* Dordrecht, The Netherlands, Reidel Publishing Company. pp. 269–305.

Keesmann, S., Noseck, U., Buhmann, D., Fein, E. & Schneider, A. (2005) *Modellrechnungen zur Langzeitsicherheit von Endlagern in Salz- und Granitformationen (Model simulations of long-term safety of nuclear waste repositories in salt and granite formations).* Braunschweig, Germany, GRS. Final Report. BMWA-FKZ 02E9239.

Keller, S. (1998) Permafrost in der Weichsel-Kaltzeit und Langzeitprognose der hydrogeologischen Entwicklung der Umgebung von Gorleben/NW-Deutschland (Permafrost during the Weichsel cold stage and long-term prognosis of the hydrogeological development of the Gorleben area/NW Germany). *Zeitschrift für Angewandte Geologie* 48, 111–119.

Klinge, H., Boehme, J., Grissemann, C., Houben, G., Ludwig, R.-R., Rübel, A., Schelkes, K., Schildknecht, F. & Suckow, A. (2007) *Standortbeschreibung Gorleben, Teil 1: Die Hydrogeologie des Deckgebirges des Salzstocks Gorleben* (*Description of the Gorleben Site, Part 1: Hydrogeology of the Overburden of the Gorleben Salt Dome*). *Geologisches Jahrbuch*. C 71. [Online] Available in English from: http://www.bgr.bund.de/EN/Themen/Endlagerung/Downloads/Description_Gorleben_Part1_Hydrogeology_overburden_en.pdf?__blob=publicationFile&v=2 [Accessed 6th August 2013].

Klinge, H., Köthe, A., Ludwig, R.-R. & Zwirner, R. (2002) Geologie und Hydrogeologie des Deckgebirges über dem Salzstock Gorleben (Geology and hydrogeology of the overburden of the Gorleben salt dome). *Zeitschrift für Angewandte Geologie* 2, 7–15.

Kösters, E., Vogel, P. & Schelkes, K. (2000) *2D-Modellierung der paläohydrogeologischen Entwicklung des Grundwassersystems im Elberaum zwischen Burg und Boitzenburg* (*2D-Modeling of the Palaeohydrogeological Development of the Groundwater System in the Elbe Area between Burg and Boitzenburg*). Hannover, Germany, BGR.

Köthe, A., Hoffmann, N., Krull, P., Zirngast, M. & Zwirner, R. (2007) *Standortbeschreibung Gorleben, Teil 2: Die Geologie des Deck- und Nebengebirges des Salzstocks Gorleben* (*Description of the Gorleben Site, Part 2: Geology of the Overburden and Adjoining Rock of the Gorleben Salt Dome*). *Geologisches Jahrbuch*. C 72. [Online] Available in English from: http://www.bgr.bund.de/EN/Themen/Endlagerung/Downloads/Description_Gorleben_Part2_Geology-overburden-adjoining%20rock_en.pdf?__blob=publicationFile&v=1 [Accessed 6th August 2013].

Langer, A. & Schütte, H. (2002) Geologie norddeutscher Salinare (Geology of saliniferous formations in Northern Germany). *Akademie der Geowissenschaften zu Hannover* 20, 63–69.

Legrand, M.R., Lorius, C., Barkov, N.I. & Petrov, V.N. (1988) Vostok (Antarctica) ice core: Atmospheric chemistry changes over the last climatic cycle. *Atmospheric Environment* 22, 317–331.

Linke, G., Katzenberger, O. & Grün, R. (1985) Description and ESR dating of the Holsteinian interglaciation. *Quaternary Science Reviews* 4, 319–331.

Loutre, M.F. & Berger, A. (2000) Future climate changes: Are we entering an exceptionally long interglacial? *Climate Change* 46, 61–90.

Ludwig, R. & Kösters, E. (2002) Hydrogeologisches Modell Gorleben – Entwicklung bis zum paläohydrogeologischen Ansatz (Hydrogeological model Gorleben – Development up to the palaeohydrogeological approach). In: FH-DGG (ed.) *Hydrogeologische Modelle – Ein Leitfaden mit Fallbeispielen* (*Hydrogeological models – a guideline with case studies*). *German Geological Society Series*, 24, 69–77.

Mattheß, G. (1994) Die Beschaffenheit des Grundwassers (The composition of groundwater). In: Mattheß, G. (ed.) *Lehrbuch der Hydrogeologie* (*Textbook of hydrogeology*). Berlin, Germany, Gebrüder Bornträger. pp. 1–499.

Meyer. TU Braunschweig (Personal communication, 2006).

Nagra/PSI (2002) *Chemical Thermodynamic Data Base 01/01*. Parkland, FL, Universal Publishers.

Noseck, U., Brendler, V., Flügge, J., Stockmann, M., Britz, S., Lampe, M., Schikora, J. & Schneider, A. (2012) *Realistic Integration of Sorption Processes in Transport Codes for Long-Term Safety Assessments*. Braunschweig, Germany, GRS. Final Report. BMWi-FKZ 02E10518.

Noseck, U., Fahrenholz, C., Flügge, J., Fein, E. & Pröhl, G. (2009) *Impact of climate change on far-field and biosphere processes for a HLW-repository in rock salt – Scientific basis for the assessment of the long-term safety of repositories*. Braunschweig, Germany, GRS. Final Report. BMWi-FKZ 02E9954.

Parkhurst, D.L. & Appelo, C.A.J. (1999) *User's guide to PHREEQC (Version 2) – A computer program for speciation, batch-reaction, one-dimensional transport, and inverse geochemical calculations.* Denver, CO, U.S. Geological Survey Water-Resources Investigations Report 99–4259.

Rohling, E.J., Fenton, M., Jorissen, F.J., Bertrand, P., Ganssen, G. & Caulet, J.P. (1998) Magnitudes of sea-level lowstands of the past 500,000 years. *Nature* 394, 162–165.

Streif, H. (2004) Sedimentary record of pleistocene and holocene marine inundations along the North Sea coast of Lower Saxony, Germany. *Quaternary International*, 112, 3–28.

Streif, H. (2007) *Das Quartär in Niedersachsen und benachbarten Gebieten – Gliederung, geologische Prozesse, Ablagerungen und Landschaftsformen (The Quaternary in Lower Saxony and Neighbouring Areas – Stratification, Geologic Processes, Sediments and Landscape Morphology).* [Online] Retrieved from: http://www.lbeg.niedersachsen.de/download/1015/Quartaerstratigraphie_von_Niedersachsen_und_benachbarten_Gebieten_2007_.pdf [Accessed 6th August 2013].

Suter, D., Biehler, D., Blaser, P. & Hollmann, A. (1998) Derivation of a sorption data set for the Gorleben overburden. In: Roth, A. (ed.) *Proceedings of DisTec 98, International Conference on Radioactive Waste Disposal.* Hamburg, Germany, Kontec Ges. für Techn. Kommunikation. pp. 581–584.

Williams, R.S. & Hall, D.K. (1993) Glaciers. In: Gurney, R.J., Foster, J.L., & Parkinson, C.L. (eds.) *Atlas of satellite observations related to global change.* Cambridge, UK, Cambridge University Press. pp. 401–422.

Chapter 12

Identification and management of strategic groundwater bodies for emergency situations in Bratislava District, Slovak Republic

J. Michalko[1], J. Kordík[1], D. Bodiš[1], P. Malík[1],
R. Černák[1], F. Bottlik[1], P. Veis[1,2] & Z. Grolmusová[1,2]
[1]State Geological Institute of Dionyz Stur, Department of Hydrogeology,
Geothermal Energy and Environmental Geochemistry, Bratislava,
Slovakia
[2]Department of Experimental Physics, Faculty of Mathematics, Physics
and Informatics, Comenius University in Bratislava, Bratislava, Slovakia

ABSTRACT

Drinking water of sufficient quality and quantity is required to maintain human health and well-being. The drinking-water resources are potentially endangered in many urbanised environments affecting the quality of life and human health. In 2010 Slovakia initiated a project aimed at the identification and management of strategic groundwater bodies within its capital district, Bratislava. Existing data and information were identified and compiled and a basic methodology for selection of alternative water resources was prepared. The results of the project are given in the paper concerning the Zitny ostrov area as a pilot case study. The basic criteria for the identification of alternative resources of drinking water are the available quantity and satisfactory quality indicators and low groundwater vulnerability. In terms of the Zitny ostrov area, the change in groundwater chemistry and quality with depth plays the most important role in the identification of alternative groundwater resources. Generally, the groundwater quality is negatively affected up to a depth of 25 m due to specific natural conditions and anthropogenic pressures. In contrast, deeper horizons are characterised by groundwater with very good quality. It can be concluded that alternative resources of potable groundwater fitting the quantitative and qualitative criteria in the Zitny ostrov area represent the aquifer horizons with depths greater than 90 m.

12.1 INTRODUCTION

Sudden changes in groundwater quantity and quality can potentially cause serious problems in drinking-water supply. The main task is therefore to develop different (contemporary) scenarios of these changes that focus mainly on the search for alternative potable groundwater sources in terms of hydrogeological and hydrogeochemical criteria. In order to identify suitable alternative resources, stable isotope studies are carried out, as well as gathering of information and making it available through GIS, and economic issues, management and other aspects are considered. Floods or droughts caused by global climate change, geohazards and sudden contamination events caused by nuclear or industrial accidents can (de Melo *et al.*, 2008) to a large extent affect the accessibility of drinking-water supply. There is a demand to prevent the effects of such scenarios (Capelli *et al.*, 2001) through identification and characterisation of 'strategic sources of groundwater' and assembled in functioning GIS as a background for groundwater-quality management (Canter *et al.*, 1994). Sheriff *et al.*, 1996; Scawthorn *et al.*, 2000; Lowry *et al.*, 2003; Verjus, 2003; Perfler *et al.*, 2007; Schwecke *et al.*, 2008; Carrera-Hernandez and Gaskin (2009) and others discuss the relations of alternative (strategic) groundwater sources for the supply of irrigation needs, economic aspects and conflicts in water-management plans. Nowadays Bratislava, the capital city of Slovakia with about 500 000 inhabitants, is failing to cope with important issues, such as sudden quantitative and qualitative changes, as well as other kinds of groundwater status changes, thus endangering the city's groundwater supply. The Bratislava metropolitan area is supplied with high-quality drinking water, sourced from Quaternary sediments of the Danube River – Žitný ostrov area. However, identification of strategic resources is needed for future groundwater management. This is the main motivation for the project entitled, 'Eco-technology for selection and assessment of alternative sources of drinking groundwater, pilot area Bratislava Region' that is implemented by State Geological Institute of Dionyz Stur (SGUDS). The project is supported with financial help of the European Regional Development Fund (ERDF) and Agency of the Ministry of Education, Science, Research and Sport of the Slovak Republic for the EU structural funds (ASFEU). The core output of the project is to develop the methodology and technology for the selection of alternative sources and assessment of groundwater quality for drinking purposes. Bratislava region was selected as a pilot area due to its quite complex geological and hydrogeological setup, where a relatively small area offers suitable conditions for the development of such methodology. The partial results of the project are presented here with the focus on the aquifer of Danube River Quaternary sediments in the Bratislava metropolitan area (Figure 12.1).

12.2 MATERIAL AND METHODS

Documentary material was obtained on the basis of a detailed study of hydrogeological and hydrogeochemical archive data (more than 1 000 scientific reports, technical publications and manuscripts over the period of the past 50 years, which give information on approximately 2 000 water resources in the territory) of the existing groundwater resources (archive of State Geological Institute of Dionyz Stur). The hydrogeological database of existing natural and man made water sources as well as the database of boreholes (wells) and database of springs – not included in the pilot area) contains basic information about specific quantitative parameters (yield, utilisation status, exploitation amount, etc.), characteristics of the rock environment along the groundwater circulation, natural recharge conditions, etc. In order to provide the methodology for the selection of alternative water resources the study

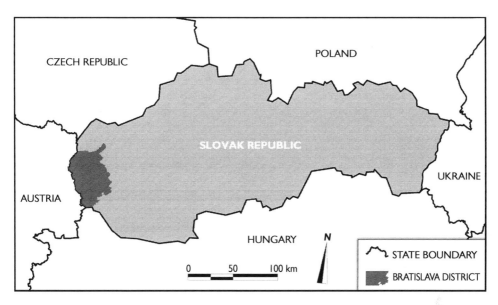

Figure 12.1 Area of interest.

entailed the basic or detailed hydrogeological investigation, providing a source of drinking water, use and protection of groundwater, groundwater monitoring and remediation. The database comprises a total of 1 968 boreholes (or wells). As an additional source of information the hydrological data from the archives of the Slovak Hydrometeorological Institute (SHI) were processed. The SHI data comprise information about the use of water resources for water-supply purposes, water withdrawal and customer data.

The hydrogeochemical database (springs, boreholes, wells, drainages) includes basic information about a specific qualitative status and chemical composition. It contains a total of 1 054 records on the chemical composition of groundwater resources since 1991 mostly (from 932 sampling sites). The database contains basic physical and chemical parameters of water quality (water temperature, pH, Eh, conductivity, major cations and anions) and the concentrations of trace elements and organic components to a lesser extent (most of the data are cited in Hanzel, 1993; Rapant *et al.*, 1996; Vrana *et al.*, 1987).

Based on the archive study of hydrogeological and hydrogeochemical data, potential potable groundwater resources were selected for further study (sampling and analytical processes in 2010–2011; a total of 28 analyses).

The key factors taken into account when identifying new strategic resources of drinking water (from the data obtained) include: access to water resource and its safety; potential sources of contamination (land use) and water protection; quality requirements; treatment of drinking water in terms of national (legislative) criteria; the cost of construction and operation of the source; and the average yield of the source with respect to its usability.

12.3 CASE STUDY – ZITNY OSTROV AREA

12.3.1 Geology

The Great Pannonian Basin, including the important water management area Zitny ostrov, has a dish-shaped brachysynclinal structure confined to the edges by faults (Figure 12.2). The entire study area lies in the Gabcikovo Depression (central part of the Great Pannonian Basin), which started to subside in the Badenian period and the entire development process continued up to the end of the Pliocene period. The main Neogene filling of the brachysyncline–shaped Pannonian Basin is brackish and fresh-water sediments of the Upper Miocene (Pannonian – Pontian) and Pliocene (Dacian, Rumanian). Marine and brackish sediments of Badenian and Sarmatian (Miocene), or also buried Miocene volcanic centres are to be found beneath. The thickness of Neogene sediments (mainly clays and sands, deposited in lacustrinal and deltaic environments) in the centre of the Gabčíkovo Depression reaches nearly 3 500 m (Janáček, 1967).

Geological development of the area during the Quaternary period occurred as a result of complicated neo-tectonic movements of partial morpho-tectonic structures of the Great Pannonian Basin and mountain range of the Western Carpathians and

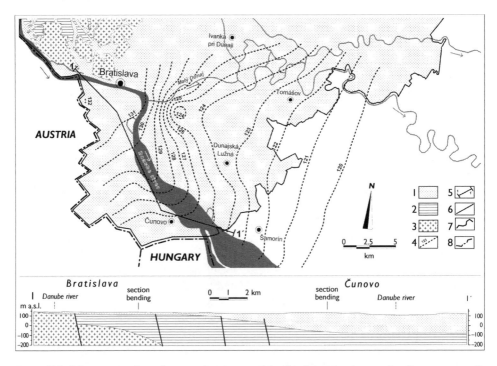

Figure 12.2 Schematic geological map and piezometric levels of case-study area. 1 – Quaternary sediments; 2 – Neogene sediments; 3 – Crystalline rocks; 4 – Groundwater level contour in Quaternary aquifer on 7.6.2000 [m amsl]; 5 – Line of Section; 6 – Faults (in section); 7 – Bratislava Self-Governing Region; 8 – State borders.

relevant forming and distribution of the sediments of Danube and its tributaries: Cierna voda, Dudvah, Vah. Among the Quaternary sediments the most important and practically dominant (from the point of view of genesis, volume, areal extension, stratigraphy and occurrence) are fluvial accumulations of Quaternary watercourses (Lower Pleistocene – Holocene). The subsiding basins in the central part of the Great Pannonian Basin were filled up with fluvial sediments, also during the Quaternary period. Quaternary fill of the basin in the Zitny ostrov area consists of three more remarkable complexes of Lower Pleistocene; Middle and Upper Pleistocene; and Holocene. Accumulations of Lower Pleistocene in superposition development were found only in the central part of the Great Pannonian Basin, where their base is at a depth of up to 500 m and thickness can reach up to 340 m (Császár *et al.*, 2000; Scharek *et al.*, 2000). Except for the central part of the Gabcikovo Depression these sediments are in discordant position on subjacent members of Upper Neogene and towards to edges of the depression their thickness is lower, up to some 10 m. They are not present on the surface. Geological development in the Middle and Upper Pleistocene is characterised by extensive fluvial sedimentation of Danube River and its Carpathian's tributaries. The development of the central Gabcikovo Depression basin is followed by synsedimentary subsidence with gradual incorporation of more stable, respectively less intensively subsiding marginal parts. For mentioned period is typical deposition of middle complex sediments named as Danubian gravel series (Janáček, 1967; 1969). This complex is created by Middle up to Upper Pleistocene fluvial sediments of the Danube River and the Vah River. In the central part of the depression its thickness reaches up to 160 m and on the margins towards the uplands it is reduced to a thickness of 50 m to 30 m. The complex consists of middle-grained to coarse-grained gravels, sandy gravels, sands and sporadic thick interglacial layers of clay and soils with fossil fauna (Pristaš *et al.*, 1996). Holocene series of upper complex (in broader sense flood facies) create lithofacialy variable laterally changing flood cover on Upper Pleistocene sandy gravels of the Danube River, the Vah River and their tributaries. They form a substantial part of the Zitny ostrov surface area and are represented by loamy and loam-sandy flood sediments. Their thickness increases from the core of Zitny ostrov area towards the main watercourses up to 3.5 m to 5 m. Flood sediments form a substantial part of the Zitny ostrov surface area which is diversified by a dense network of oxbow lakes at different stages of evolution.

12.3.1.1 Hydrogeological settings

Two different aquifer types (Figure 12.2) are found in the area: gravelly Quaternary aquifer of the main Danubian floodplain and mostly sandy Neogene aquifers beneath (Pliocene – Pannonian and Dacian by their stratigraphy). However, there was no outcropping of Neogene aquifers in the entire studied area. In terms of groundwater quantities, a dominant position belongs to the Danubian fluvial accumulations (gravels and sandy gravels), formed in the stratigraphic interval of the Lower Pleistocene to Holocene. The greatest accumulation (up to 500 m of thickness) of Quaternary gravels was documented in the central part of the Great Pannonian Basin – in the Gabčíkovo Depression, where they were deposited (for Quaternary) in an atypical superpositional development, starting from a locally well-preserved transitional fluvio-limnic layers of Upper Pliocene/Lower Pleistocene). This means, that in the centre of the Gabčíkovo Depression, Quaternary sedimentation continued uninterrupted since the Pliocene, while in the Bratislava area Quaternary terraces are well-preserved. In the studied area, the average transmissivity value of $4.07 \cdot 10^{-2}$ m²/s (3 516 m²/d) is the reason for the extremely high degree of permeability here. The mean hydraulic conductivity values are reported as $5.30 \cdot 10^{-3}$ m/s (458 m/d; Benková et al., 2005). Groundwater circulation in Neogene sediments is bound to the sands or sandstones of Pannonian and Pontian. In the depth interval between 45 m and 280 m the main Neogene aquifer is represented mainly by lenses of sands and sandy gravels of Rumanian and Dacian (Benková et al., 2005). The same author reported the average transmissivity and hydraulic conductivity values for Neogene aquifers as $1.18 \cdot 10^{-3}$ m²/s (102 m²/d) and $1.01 \cdot 10^{-4}$ m/s (8.7 m/d), respectively.

The general direction of groundwater flow in the main (Quaternary) aquifer is predominantly parallel to the main streams in the area (Danube River and its branch Malý Dunaj River). Local variations in flow direction are found around large groundwater sources with intensive pumping and around hydraulic curtains surrounding the refinery near Bratislava, built in the 1970s to protect groundwater. Here, the flow direction usually varies from the Danube to the centre of artificial groundwater depression.

The average groundwater level within the studied area ranges from 1.5 m to 6.7 m below the ground surface. Surface water of the Danube River feeds groundwater resources at all water stages, with the exception in the north-western part of the inundation area during the low water stage – where, under specific circumstances, the Danube starts to be a draining river (Benková et al., 2005). Groundwater level fluctuation is also affected by other factors, such as drainage and irrigation systems, or by water-level regulations in seepage canals. Close to the Danube River, the groundwater-level fluctuation follows the water-level fluctuation in the Danube almost immediately. Precipitation affects the groundwater regime in the study area only indirectly, especially in the summer period, when the increased discharge in the Danube (due to precipitation) and subsequent water-level rise in the area consequently increase groundwater levels. This happens in different time delays, depending on the distance to the surface flow (Benková et al., 2005). In many cases, the Danube water level is dependent on precipitation events in distant areas, which may not have reached the described region, as well as on snow melting in the Alps. At distances further away from the Danube, the groundwater-level fluctuation also depends on local seasonal effects – especially on the relationship between precipitation (including local snowmelt), evaporation from the soil surface and vegetation evapotranspiration. A network of irrigation canals and drainage systems provides a stabilising effect on groundwater levels.

Groundwater circulation in less permeable Pliocene aquifers is more stable and less intense. Due to its isolation by less permeable horizons, several independent partial circuits with different groundwater flow directions may occur here, although the intensity of circulation, compared to the Quaternary aquifers, can already be orders of magnitude lower. Lack of observation objects for the deeper Neogene aquifers is the reason why gradients or flow directions of deeper groundwater circulation cannot be sufficiently quantified.

12.3.1.2 Hydrogeochemical characteristics

Groundwater chemical composition in the area depends mainly on the:

- chemical composition of Danube River water (initial water) a water-level fluctuation of the Danube River with the phase shift;
- flow-path length and geochemical processes taking place after the Danube river water enters the fluvial sediments;
- position of groundwater infiltration from the river bed and the time of infiltration;
- impact of point and diffuse sources of contamination from industrial and agricultural activities;
- natural source of iron and manganese in the rock environment (in the areas of Fe, Mn accumulation a reducing environment is formed in the aquifer, while increasing their content in groundwater);
- calcium content in Quaternary sediments.

Groundwater of fluvial sediments is characterised by the strong variability in chemical composition (Table 12.1). Results are divided according to depth of groundwater sampling into samples collected from the top of the aquifer (up to 25 m) and samples taken from greater depths (over 25 m). In both groups, the dominant presence of Ca^{2+} was detected. Relatively large number of samples characterises increase contents of Mg^{2+}, Na^+ and K^+. Among anions, HCO_3^- or SO_4^{2-} dominates in groundwater of both groups. The dominant presence of chloride was found in 4 samples. Values of total dissolved solids (TDS) of the upper part of the hydrogeological structure range from 108 mg/l to 1 372 mg/l, but less significant variability of the TDS values and concentrations of basic analytical components was detected in deeper groundwaters (TDS values vary from 138 mg/l to 538 mg/l). Groundwaters discharging from the top of the aquifer are significantly mineralised due to specific environmental conditions and anthropogenic influence, compared to groundwaters of deeper circulation. Anthropogenic impact having a significant influence on the chemical composition of shallow groundwaters is mainly indicated by higher levels of nitrates, chlorides and sulphates. In the deeper part of the hydrogeological structure groundwater with good quality characteristics was identified.

12.3.1.3 Overview of groundwater exploitation

The Slovak Hydrometeorological Institute monitors groundwater exploitation in areal units. An overview of the overall groundwater exploitation in the Bratislava metropolitan area, aimed at regional water supply, where the total withdrawal for the period 1995–2009 is shown, is documented in Table 12.2 (Benková et al., 2005; Čaučík et al., 2010).

Groundwater exploitation is primarily aimed at withdrawals for public water supply, while for the food industry and other industries it represents less than 20%.

12.3.1.4 Strategic groundwater resources for drinking purposes

The basic criteria for the identification of alternative resources of drinking groundwater are the available quantity and satisfactory quality indicators and low vulnerability. In terms of the Zitny ostrov area, the change of groundwater chemistry and quality with depth in the profile of Quaternary and Neogene sediments plays the most important

Table 12.1 Statistical assessment of selected groundwater chemical parameters in the Žitný ostrov area and Danube right bank area (within the Bratislava self-governing region).

	n	Average	Median	Standard deviation	Minimum	Maximum
<25 m						
TDS	194	590	556	263	108	1 372
Na^+	193	22.3	17.5	16.7	1	76.3
K^+	194	7.84	4.24	9.9	1.2	65.5
Ca^{2+}	194	86.7	85.0	53.1	1.2	266
Mg^{2+}	194	31.3	25.8	20.7	1.7	132
Fe	193	0.473	0.48	0.513	0.1	4.49
Mn	192	0.350	0.25	0.229	0.1	1.78
NH_4^+	191	0.359	0.25	0.270	0.1	1.9
Cl^-	194	43.9	34.0	31.8	1.99	173
SO_4^{2-}	194	99.7	75.2	80.2	0.5	348
NO_2^-	44	0.454	0.5	0.473	0.05	3.2
NO_3^-	194	38.4	15.8	52.0	0.13	348
PO_4^{3-}	187	0.531	0.5	0.905	0.05	11.75
HCO_3^-	194	246	281	140	2.14	530
>25 m						
TDS	15	363	363	89.2	138	538
Na^+	15	10.5	8.74	6.44	1.4	25.6
K^+	15	2.33	2.26	1.06	0.96	5.1
Ca^{2+}	15	59.7	63.2	34.0	6.2	139
Mg^{2+}	15	16.0	14.7	7.94	2.67	35.2
Fe	15	0.555	0.5	0.411	0.1	1.87
Mn	15	0.403	0.4	0.189	0.18	0.83
NH_4^+	11	0.439	0.5	0.207	0.1	0.9
Cl^-	15	16.3	14.5	12.2	1	54.6
SO_4^{2-}	15	44.8	29.5	39.7	7.3	166
NO_2^-	11	0.364	0.3	0.273	0.1	1
NO_3^-	15	6.574	6.5	7.94	0.25	31.8
PO_4^{3-}	9	0.467	0.5	0.1	0.2	0.5
HCO_3^-	15	203	226	78.0	22	293

Table 12.2 Overall usage of groundwater in the Bratislava region (after data of Slovak Hydrometeorological Institute; Čaučík *et al.*, 2010).

Total withdrawal (l/s)								
Year	1995	1996	1997	1998	1999	2000	2001	2002
Bratislava surroundings	5 654	5 406	5 224	5 088	4 930	4 826	4 531	4 441
Year	2003	2004	2005	2006	2007	2008	2009	
Bratislava surroundings	4 810	4 228	4 153	4 214	4 296	4 188	4 201	

role in identification of alternative groundwater resources. From a qualitative point of view this fact is documented in Figure 12.3, which shows the variation of nitrate concentrations with depth as an indicator of anthropogenic influence from diffuse pollution sources but also from point-source contamination. Significant reduction of nitrates with depth in groundwater is obvious (Figure 12.3) and from a depth of about 25 m the values practically do not exceed the permissible concentration (50 mg/l).

Following changes of the total dissolved solids (TDS) in groundwater with depth (Figure 12.4), two functionalities are obvious. The first are high levels of TDS caused by anthropogenic activities up to a depth of approximately 25 m. Secondly, the increase in TDS with depth is characteristic for the Neogene aquifers. In this case, the increase in TDS is conditioned by ion-exchange processes and mainly associated with an increase in the content of sodium and bicarbonates, accompanied by a chemical process to produce $NaHCO_3$.

The entire hydrogeological region of Neogene sediments characterises the high transmissivity (Figure 12.5) – median of transmissivity coefficient T is $1.2 \cdot 10^{-3}$ m²/s. (Benková *et al.*, 2005). The specific capacity of wells q ranges from 0.02 l/s · m to 21.98 l/s · m (average value of q is 3.00 l/s · m).

The Danube River with a length of 2 860 km (the second longest river in Europe) and average discharge before emptying into the Black Sea via the Danube Delta of 6 500 m³/s is considered the primary source of groundwater in the area. In its upper reach (from its spring source to Bratislava, 1 000 km) the physicochemical composition

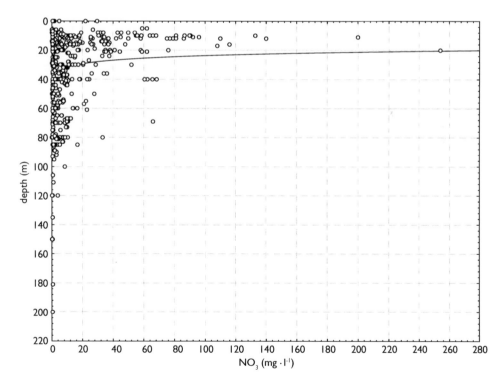

Figure 12.3 Variation of nitrate concentrations with depth.

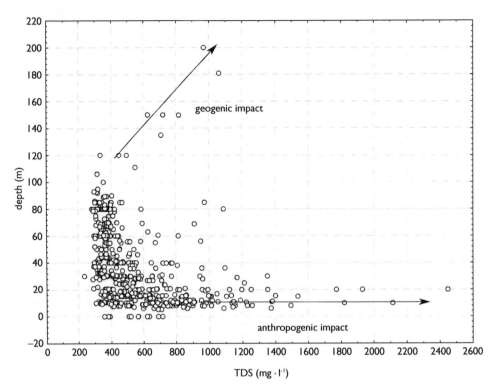

Figure 12.4 Variation of TDS with depth.

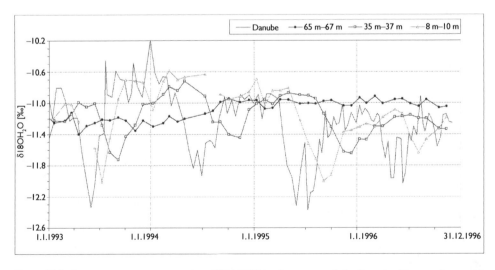

Figure 12.5 Oxygen isotope composition (VSMOW) in the water of the Danube River and ground-water in different depths of piezometer 6030, Čunovo, Danube right bank (based on data by Michalko *et al.*, 1997).

Table 12.3 Isotope composition of Danube River water, Morava River (local river) and precipitation (data from Rank et al., 2012 and Michalko, 1998).

Locality	Source	Sample type	Period	$\delta^{18}O$ [‰]
Vienna	Danube River	Monthly grab	1976–1985	−11.75
			1996–2005	−11.22
Angern	Morava River	Monthly grab	1976–1985	−9.88
			1996–2005	−9.33
Bratislava 286 [m amsl]	Precipitation	Monthly average	1988–1997	−8.70
Topol'níky 113 [m amsl]	Precipitation	Monthly average	1988–1997	−9.19

of Danube water is controlled by its right-side tributaries mainly through the dissolution of carbonate complexes of the Alps (Pawellek et al., 2002). Quaternary fluvial sediments of the Danube in the Great Pannonian Basin (also included in the case-study region) represent one of the largest freshwater reservoirs in Europe. Due to its origin the Danube water is significantly different in its isotope composition from local meteoric waters and their derivates – tab. 12.3 (Pawellek et al., 2002; Rank et al., 2012; Michalko 1998).

The fact that the groundwater of the case study area originates from the Danube River is also evidenced by data on the isotopic composition ($\delta^{18}O$) of the Danube River monitored in the years 1983–1998 in Bratislava as well as by knowledge of the isotopic composition of groundwater and surface water in the Bratislava metropolitan area (Kantor et al., 1989). Similarly, the data on isotopic composition ($\delta^{18}O$) of groundwater of the Zitny ostrov area gathered through monitoring in 1991–1996 from 30 piezometers during the filling of the Gabcikovo Dam support this fact (Michalko et al., 1997; Michalko, 1998). Based on these data it can be assumed that the current Danube River water infiltrates the shallower aquifer horizons, particularly in preferred areas. Rodák et al. (1995) in their study on the Kalinkovo area assume intrusion of the Danube River water in shallow horizons up to 4 km to 6 km within two to three years. Similarly, it is possible to prove (Figure 12.5) relatively rapid infiltration of the Danube River water (red) in the 8 m to 10 m zone (light blue) and only slow partial water exchange at a depth of 35 m to 37 m (blue). The effect of the Danube River water on the isotopic composition of groundwater at a depth of about 67 m (dark blue) is minimal. It can be assumed that the water of the Danube River currently has a minimal impact on groundwater of deeper horizons of the hydrogeological structure. Consequently, the higher residence time and better water quality is expected in deeper horizons of the Quaternary and Neogene sediments.

12.4 CONCLUSION

Sudden unexpected changes in groundwater quality and quantity can potentially cause serious problems in drinking-water supply to inhabitants. Identification of strategic resources of potable groundwater which are available in emergency situations should be an essential part of water-management plans in any country. This paper focuses attention on identification of strategic drinking-water sources waters in the pilot area (Žitný ostrov and right bank of the Danube River) within the Bratislava self-governing region.

In the pilot area, a substantial deterioration in groundwater quality due to specific natural conditions (natural source of Fe and Mn) and anthropogenic activities (industrial and agricultural) was observed in the upper part of the hydrogeological structure. Anthropogenic impact on groundwater is mainly indicated by moderate to high increases in the levels of chlorides, sulphates, and nitrates). Deeper parts of the hydrogeological structure (depths greater than 25 m) are characterised by groundwater of very good quality and in terms of current as well as strategic use of groundwater for drinking purposes they have crucial significance. Groundwater of deeper aquifer horizons originates from the Danube River; however, it can be assumed that the recent impact of the Danube River on deeper groundwater quality and chemistry is minimal.

Although the results showed a good opportunity to use the Zitny ostrov and the right-bank area of the Danube River as a strategic drinking-water supply in emergency situations, the main limitations of the investigation were the lack of quantitative and qualitative data from (perspective) deeper horizons of the hydrogeological structure and the failure to address the evaluation of long-term observations of chemical composition and quality of groundwater monitoring sites. Therefore, more attention should be given to these issues in the future along with the close cooperation of organisations dealing with water management in Slovakia.

ACKNOWLEDGEMENTS

The authors gratefully acknowledge the support of the European Regional Development Fund and the Agency of the Ministry of Education, Science, Research and Sport of the Slovak Republic for the EU structural funds (ASFEU) through the project 'Ekotechnológia vyhl'adania a hodnotenia náhradných zdrojov pitných podzemných vôd, pilotné územie BSK [Ecotechnology of identifying and assessment of alternative drinking groundwater resources]' (ITMS: 26240220003).

REFERENCES

Benková, K., Bodiš, D., Nagy, A., Maglay, J., Švasta, J., Černák, R., Marcin, D. & Kováčová, E. (2005) Základná hydrogeologická a hydrogeochemická mapa Žitného ostrova a pravobrežia Dunaja v mierke 1 : 50 000 [Basic hydrogeological and hydrogeochemical map of the Žitný ostrov and Danube right bank area at a scale 1:50 000, in Slovak]. Manuscript – Archive of Geofond ŠGÚDŠ Bratislava, Slovakia.

Canter, L.W., Chowdhury, A.K.M.M. & Vieux, B.E. (1994) Geographic information systems: A tool for strategic ground water quality management. Journal of Environmental Planning & Management 37, 251–266.

Capelli, G., Salvati, M. & Petitta, M. (2001) Strategic groundwater resources in Northern Latium volcanic complexes (Italy): Identification criteria and purposeful management. IAHS Press 272, 411–416.

Carrera-Hernandez, J.J. & Gaskin, S.J. (2009) Water management in the Basin of Mexico: Current state and alternative scenarios. *Hydrogeology Journal* 17, 1483–1494.

Császár, G., Pistotnik, J., Pristaš, J., Elečko, M., Konečný, V., Vass, D. & Vozár, J. (2000) Surface geological map. In: Császár, G. (ed.) *Danube Region Environmental Geology Programme DANREG Explanatory Notes, Jb. Geol. B. – A.,* Volume 4. Wien. pp. 421–455.

Čaučík, P., Mihálik, F., Leitmann, Š., Gavurník, J., Sopková, M., Možiešiková, K., Stojkovová, M., Molnár, L., Bodácz, B., Lehotová, D. & Juráčková, D. (2010) *Vodohospodárska bilancia SR. Vodohospodárska bilancia množstva podzemnej vody za rok 2009* [Water-Management Balance of the Slovak Republic. Water-Management Balance of Groundwater Amounts for 2009, in Slovak]. Bratislava, Slovakia, Slovenský hydrometeorologický ústav.

de Melo, M.T.C., Fernandes, J., Midoes, C., Amaral, H., Almeida, C.C., da Silva, M.A.M. & Mendonca, J.J. (2008) Identification and management of strategic groundwater bodies for emergency situations in Portugal (IMAGES). In: *Abstracts – 33rd International Geological Congress – Congres Geologigue International Resumes.*

Hanzel, V., Vrana, K. & Cimborová, S. (1993) *Podzemné vody západných svahov Devínskych a Pezinských Karpát, čiastková záverečná správa* [Groundwater of the Devínske and Pezinské Karpaty Mts. – western slopes, in Slovak]. Manuscript – Archive of Geofond ŠGÚDŠ Bratislava, Slovakia.

Janáček, J. (1967) *Výskum tektoniky J časti Podunajskej nížiny s ohľadom na výstavbu VD Dunaj, záverečná správa* [Investigation of tectonics in the southern part of the Podunajska nizina Basin concerning the Gabcikovo water dam construction, in Slovak]. Manuscript – Archive of Geofond ŠGÚDŠ Bratislava, Slovakia.

Janáček, J. (1969) Nové stratigrafické poznatky o pliocénní výplni centrální části Podunajské nížiny [New stratigraphic knowledge about Pliocene sediments of the central part of the Podunajska nizina Basin, in Czech]. *Geologické Práce, Správy, ŠGÚDŠ Bratislava* 50, 113–131.

Kantor, J., Ďurkovičová, J. & Michalko, J. (1989) *Izotopový výskum hydrogenetických procesov – II. časť'* [Isotope investigation of hydrogenetic processes – II. Part, in Slovak]. Manuscript – Archive of Geofond ŠGÚDŠ Bratislava, Slovakia.

Lowry, T.S., Bright, J.C., Close, M.E., Robb, C.A., White, P.A. & Cameron, S. (2003) Management gaps analysis; a case study of groundwater resource management in New Zealand. *International Journal of Water Resources Development* 19, 579–592.

Michalko, J., Ferenčíková, E., Rúčka, I. & Kovářová, A. (1997) *Správa o výsledkoch meraní izotopového zloženia kyslíku povrchových a podzemných vôd v oblasti vodného diela Gabčíkovo a Žitného ostrova v roku 1996* [Isotopic composition of oxygen in surface and ground water in the Gabcikovo area and Zitny ostrov territory – Report on measurement results, in Slovak]. Manuscript – Archive of Geofond ŠGÚDŠ Bratislava, Slovakia.

Michalko, J. (1998) *Izotopová charakteristika podzemných vôd Slovenska, kandidátska dizertačná práca časť'* [Isotopic characterization of the Slovak groundwater, dissertation thesis, in Slovak]. Slovak Academy of Science, Bratislava, Slovakia.

Pawellek, F., Frauenstein, F. & Veizer, J. (2002) Hydrochemistry and isotope geochemistry of the upper Danube River. *Geochimica et Cosmochimica Acta* 66, 3839–3854.

Perfler, R., Unterwainig, M., Mayr, E. & Neunteufel, R. (2007) The security and quality of drinking water supply in Austria – Factors, present requirements and initiatives. *Österreichische Wasser- und Abfallwirtschaft,* Springer-Verlag Wien, 59, 125–130.

Pristaš, J., Halouzka, R., Horniš, J., Elečko, M., Konečný, V., Lexa, J., Nagy, A., Vass, D. & Vozár, J. (1996) *Povrchová geologická mapa (Podunajsko – Danreg) 1:100 000, M-33-143* [Surface geological map (Podunajsko – Danreg) at a scale 1:100 000, in Slovak]. In: Kováčik, M., Tkáčová, H., Caudt, J., Elečko, M., Halouzka, R., Hušták, J., Kubeš, P., Malík, P., Nagy, A., Petro, M., Pristaš, J., Rapant, S., Remšík, T., Šefara, J., Vozár, J. Podunajsko – Danreg, záverečná správa [Podunajsko – Danreg, final report, in Slovak]. Manuscript – Archive of Geofond ŠGÚDŠ Bratislava, Slovakia.

Rank, D., Papesch, W., Heiss, G. & Tesch, R. (2012) Environmental isotopes of river water in the Danube Basin. In: *Monitoring isotopes in rivers: Creation of the global network of isotopes in rivers* (GNIR). IAEA-TECDOC-1673. pp. 13–40.

Rapant, S., Vrana, K. & Bodiš, D. (1996) *Geochemical atlas of the Slovak Republic – Groundwater.* Geological Survey of the Slovak Republic, Bratislava, Slovakia.

Rodák, D., Ďurkovičová, J. & Michalko, J. (1995) *The use of stable oxygen isotopes as a conservative tracer in the infiltrated Danube river water. Gabčíkovo part of the hydroeletric power project – Environmental impact review.* Faculty of Natural sciences, Comenius University, Bratislava, Slovakia.

Scawthorn, C., Ballantyne, D.B. & Blackburn, F. (2000) Emergency water supply needs lessons from recent disasters. *Water Supply,* IWA Publishing, London, United Kingdom, 18, 69–77.

Scharek, P., Herrmann, P., Kaiser, M. & Pristaš, J. (2000) Map of genetic types and thickness of quaternary sediments. In: Császár, G. (ed.) *Danube Region Environmental Geology Programme DANREG Explanatory Notes.* Jb. Geol. B. – A., Volume 4. Wien. pp. 447–455.

Schwecke, M., Simons, B., Maheshwari, B. & Ramsay, G. (2008) Integrating alternative water sources in urbanized environments. In: *Proceedings of the 2nd International Conference on Sustainable Irrigation Management, Technologies and Policies, Sustainable Irrigation. WIT Transactions on Ecology and the Environment 112.* United Kingdom, WIT Press. pp. 351–359.

Sheriff, J.D., Lawson, J.D. & Askew, T.E.A. (1996) Strategic resource development options in England and Wales. *Chartered Institution of Water and Environmental Management* 10, 160–169.

Verjus, P. (2003) *Albian-Neocomien Underground Water Resources. How to Safeguard and Manage an Emergency Strategic Resource for Drinking Water.* Paris, France, Societe Hydrotechnique de France. pp. 51–56.

Vrana, K., Pospiechová, O. & Vyskočil, P. (1987) *Mapa kvalitatívnych vlastností podzemných vôd územia Veľkej Bratislavy (časť' sever), mierka 1 : 50 000 [Map of groundwater quality of the Bratislava region – Northern part, scale 1 : 50 000,* in Slovak]. Manuscript – Archive of Geofond ŠGÚDŠ Bratislava, Slovakia. [Online] Available from: http://www-naweb.iaea. org/napc/ih/IHS_resources_isohis.html

Chapter 13

A model of long-term catchment-scale nitrate transport in a UK Chalk catchment

N.J.K. Howden[1], S.A. Mathias[2], M.J. Whelan[3], T.P. Burt[4] & F. Worrall[2]

[1]*Department of Civil Engineering, University of Bristol, Queen's Building, University Walk, Bristol, UK*
[2]*Department of Earth Sciences, Durham University, Durham, UK*
[3]*Natural Resources Department, School of Applied Sciences, Cranfield University, Cranfield, Bedfordshire, UK*
[4]*Department of Geography, Durham University, Durham, UK*

ABSTRACT

This paper presents a model of catchment-scale solute transport applied to the case of nitrate in a small agricultural watershed (Alton Pancras: <10 km²) in the River Piddle catchment, Dorset, UK. The solute-transport model calculates stream-nitrate response to catchment-scale nitrate loading on an annual time-step assuming: an initial baseline concentration, C_b; a nitrate load to concentration conversion factor, α; some dispersion characteristic of the catchment (hydrogeological) system, P_e; a time delay between initial catchment nitrate loading and stream concentration response, t_d; and the Mean (catchment) Travel Time (MTT). Historical land-use and management data are used to estimate the net annual loading of nitrate to the catchment between 1930 and 2007, and parameters are based on model fits to observed annual average stream-nitrate concentrations between 1981 and 2004.

A simple graphical translation of the catchment nitrate loading and river concentration data suggests an MTT of 37 years. The Mean Absolute Error (MAE) reached a minimum at $P_e = 1\,418$, and beyond this the MAE rises to a stable plateau of 0.26 mg/l. Estimates of α and C_b converge with increasing P_e. For this particular catchment the value of P_e suggests catchment-scale dispersion may be ignored, taking $P_e \to \infty$, and thus allowing model simplification. This allows catchment nitrate loading and stream response to be related by a simple linear model.

Alternative catchment nitrate-loading scenarios are considered, assuming that fertiliser inputs between 1930 and 2007 were cut by 25%, 50%, 75% and 100%. Our models show two points of interest: fertiliser inputs are only partially responsible for the stream concentration rises between 1970 and the present, but the future peak in around 2017 will be almost 50% attributable to fertiliser inputs. It is noted that rises in stream-nitrate concentration observed to date result from a combination of grassland ploughing, increasing animal inputs and fertiliser application, rather than the latter. Hence, policies that rely solely on fertiliser management address only around a third of the total inputs.

The results demonstrate that, in groundwater-dominated catchments, MTTs are of the order of several decades, even in the smallest of watersheds. Therefore diffuse pollution strategies implemented now will not have a measurable impact on the river, in this case, for almost 40 years. Further, in this particular catchment, stream-nitrate concentrations will continue to rise due to past land use and management, peaking just before 2020.

13.1 INTRODUCTION

Western Europe has long been identified as a global hot spot of fluvial nitrogen flux with some of the highest concentrations being found in UK rivers (Meybeck, 1982). Anthropogenic activity has doubled the rate at which biologically-available nitrogen enters the terrestrial biosphere when compared to pre-industrial levels (Galloway *et al.*, 2004); over the same period, global river fluxes of Dissolved Inorganic Nitrogen (DIN), of which nitrate (NO_3) is the major component, have increased sixfold (Green *et al.*, 2004). Nitrate is one of the most problematic and persistent of water contaminants (Howden and Burt, 2008). Dramatic rises in fluvial nitrate concentrations, particularly in rural catchments, are generally accepted to be the result of agricultural intensification during the latter half of the 20th century (Foster and Crease, 1974). There are significant scientific challenges in addressing the issue of diffuse nitrate pollution from agriculture for two main reasons: first, the hydrological processes linking agricultural activity and nitrate transfer from land to surface water and groundwater remain poorly understood (Royal Society, 1983; Addiscott *et al.*, 1991; Burt *et al.*, 1993; Mathias *et al.*, 2005; Howden and Burt, 2009; Burt *et al.*, 2010); and, second, the extent of diffuse pollution varies considerably as a complex function of soil type, climate, topography, hydrology, land use and management (Heathwaite *et al.*, 2005) such that managing the risk of diffuse pollution is not straightforward (Burt and Pinay, 2005). Nonetheless, it has been argued that farm management to limit the sources of diffuse nitrate pollution is the key if fluvial nitrate trends are to be reversed (Jackson *et al.*, 2008). Others have argued that nitrate could also be managed by options such as re-constructed wetlands or re-introduction of water meadows, not just by on-farm measures (Wade *et al.*, 2006; Whitehead *et al.*, 2006).

Identification of water-quality trends has often been the focus of hydrological research (e.g., Roberts and Marsh, 1987; Burt *et al.*, 1988; Betton *et al.*, 1991; Howden and Burt, 2009), but this has generally been limited by two principal factors: the ability to identify statistically significant trends from monitoring over relatively small time periods (i.e., <10 years); and, where trends can be identified, deriving a hydrological process-understanding to explain both their magnitude and likely persistence over time.

The European Union Water Framework Directive (EU-WFD, 60/2000/EC) aims to improve the ecological condition of continental water bodies using catchment-level environmental management initiatives, and promotes the analysis of temporal trends in water quality such that European water bodies exceeding established reference nutrient levels may be identified and appropriate mitigation measures prescribed. The ultimate EU-WFD target is the achievement of good ecological status for European rivers by 2015, but this may not be realistic for many lowland areas where groundwater flows are the predominant mechanism for maintaining river flows (Weatherhead and Howden, 2009). Most research on diffuse pollution has involved surface and near-surface flow paths, but there are added complexities in groundwater-dominated catchments due to the long pathways and residence times involved in nutrient transfer through the aquifer (Mathias *et al.*, 2007; Howden and Burt, 2008). One of the earliest detailed analyses of water quality trends was that by Casey and Clarke (1979), who studied rising nitrate concentrations in the River Frome, Dorset, UK, between 1965 and 1976. They used simple ANOVA techniques to show that average annual nitrate concentrations had risen by 0.107 mg/l · yr. Howden and Burt (2008, 2009) recently showed that, over the period 1977 to 2007, this rate of increase in annual mean nitrate concentration was sustained, and that 95.9% ($p < 0.001$) of the variation in annual mean concentrations in the River Frome between 1965 and 2007 could be described by a linear trend line. Whilst providing the simplest model of observed data (i.e., a linear regression using two parameters), this did not enable a hydrological process-understanding to be derived from the analyses – except that such a sustained rise in concentration over the 42 years was most striking. This was attributed to the fact that, in the Frome catchment, between 75% and 95% of river flows originate from groundwater (Howden, 2004): the groundwater thus acts as a large storage over time, where successive inputs (recharge waters) travel slowly through the system to emerge many years (or even decades) later. In terms of solute transport from land to river, this leads to two characteristics in the hydrological response: a long time delay between cause (i.e. change in N-loading to the catchment) and effect (i.e. fluvial concentration increase); and chemical buffering, due to the homogenising effect of the large groundwater storage (Howden *et al.*, 2010). Howden and Burt (2008) used data from 34 sites throughout the River Frome and neighbouring River Piddle catchments from 1977 to 2007, and specific observations from the headwater areas, to suggest the long-term response could be approximated by a form of solute breakthrough curve; this was applied to all 34 data sets, showing both an improved fit to the data and, crucially, adding the potential for a process-interpretation of the trends.

Here we use observations from the headwaters of the River Piddle at Alton Pancras (<10 km² watershed), together with estimated N-loadings from watershed land use and land management between 1930 and 2007 to identify key characteristics of both the nitrate breakthrough, and its relation to historical land-use and land-management patterns. We develop a predictive impulse-response function relating the estimates of past annual catchment N-loading to future annual average nitrate concentrations in the river. Specifically, we consider the following questions:

1   How do observed rises in annual average fluvial nitrate concentration relate to present and past land-use and land-management practices, and can the time delay between changing land-use and/or land-management practice and fluvial nitrate concentration response be estimated?
2   Over this time period, what is the dispersive effect of solute transport through the catchment system?
3   Can we develop a simple model relating, on an annual time-step, the fluvial nitrate concentration expected from estimates of historical net catchment N-loading?
4   What implications does this have for the development of land-use and land-management policy, and for attaining water-quality targets under WFD by 2015?

13.2   CATCHMENT AND DATA

The Alton Pancras catchment (approximate area 9.21 km²) is a headwater catchment of the River Piddle in south Dorset, UK. The dominant geological formation is the

Chalk and land use is mixed, with a dominance of arable land over grassland. River flows are predominantly provided from Chalk springs and the Base-Flow Index (BFI) is approximately 0.9 (Howden, 2004; Howden and Burt, 2008). There is very little near surface runoff, and there are no sewage effluent discharges to the river upstream of the sampling location (NGR ST 69939 02450). Surface-water nitrate concentrations were measured by the Environment Agency of England and Wales (EA) between 1980 and 2004 (mg NO_3-N/l) at approximately monthly intervals (see Figure 13.2). Details of sampling methodology and analytical techniques are available from www. environment-agency.gov.uk. Land-use data were obtained from parish summary records for 1866 through to 1988, held at the National Archives at Kew Gardens, London (ref: MAF 68). Parish summaries post-1988 are held by the UK National Digital Data Archives of Datasets (UKNDAD: ref CRDA/4), but access to these data is currently restricted, so data from 1988 to 2007 were approximated from UK data using a comparison between Alton Pancras and UK data between 1930 and 1988.

13.3 APPROACH

Our approach is fourfold

1 We estimate the net catchment N-loading taking account of: atmospheric deposition, N-fixation and mineralisation; inputs from livestock; inputs from fertilisers; inputs due to enhanced mineralisation following grassland ploughing; and, N-removal due to crop uptake. This is summarised in Table 13.1.
2 A simulation model is developed, based on analytical solutions for 1D advection-dispersion equations, to transform the estimated net catchment N-load to a fluvial nitrate concentration, for annual time-steps, allowing for a time lag and attenuation.
3 The model is applied to a small watershed, where data are available to estimate inputs and average annual fluvial nitrate concentrations, and consider model parameterisation.
4 We revisit the initial catchment N-loading calculations, and modify historical fertiliser inputs to consider how these would have altered the present fluvial concentrations. The annual time-step used is calendar years so to remain consistent with the available land-use and land-management data.

The approach assumes that the solute transport process from soil to ultimate river discharge may be lumped at the catchment scale. For an annual time-step this is a reasonable assumption, particularly given the focus catchment is heavily groundwater-dominated (>90%).

13.3.1 Catchment N-loading model

The transport model calculated stream-nitrate concentration, C_n (mg/l), for the nth year from:

$$C_n = C_b + \alpha M_n \tag{13.1}$$

Table 13.1 Components of the catchment nitrogen-loading model.

Mass balance

$$L = L_B + \left(I_{fert} + I_{animal} + I_m - U\right)$$

where:

L_B is the contribution from atmospheric deposition, N fixation and mineralisation. Taken as 60 kg N/ha · yr

Fertiliser inputs

$$I_{fert} = I_{arable} f_{arable} + I_{grass} f_{grass}$$

where:

I_{arable} and I_{grass} are the average N-fertiliser application rates (kg N/ha · yr) for arable and grass land, respectively

f_{arable} and f_{grass} are the catchment area fractions under arable land (including temporary grassland) and permanent grassland (including rough grazing), respectively

Animal inputs

$$I_{animal} = \frac{1}{A} \sum_{i=1}^{5} N_i E_i$$

where:

N_i is the number of animals of type i in the catchment
E_i is the N-excretion per animal (kg N/head · yr)
$i = 5$ (cattle, sheep, pigs, poultry and horses) and their excretion rates were taken as 85.7 kg N/head · yr, 6.9 kg N/head · yr, 8.8 kg N/head · yr, 0.44 kg N/head · yr and 50 kg N/head · yr

Enhanced mineralisation due to ploughing

$$I_m = \left(M_A + \left(M_P - M_A\right)\exp\left[-kt\right]\right)\left(1 - \exp - \exp\left[-kt\right]\right)$$

where:

M_A is the equilibrium mass of N under the arable regime (6 733 kg N/ha)
M_P is the equilibrium mass of N under the permanent grassland regime (10 687 kg N/ha)
k is a first-order rate constant (0.132/yr). Parameter values were taken from Whitmore *et al.* (1992)

Crop uptake

$$U = U_{min} + \frac{\left(U_{max} - U_{min}\right)\left(I_{fert} + I_{animal} + I_m\right)}{K + I_{fert} + I_{animal} + I_m}$$

where:

U (kg N/ha · yr) is crop uptake of N
U_{min} (taken as 40 kg N/ha · yr) is the minimum N uptake rate resulting from zero additional inputs over and above atmospheric deposition
N fixation and N mineralisation from soil organic N at equilibrium soil organic matter levels
U_{max} (180 kg N/ha · yr) is the maximum uptake rate (i.e., the point at which yield is no longer N limited)
I_{total} (kg N/ha · yr) is the total N input from fertilisers, animal excreta and enhanced mineralisation from ploughing up of permanent pastures
K (170 kg N/ha · yr) is a constant which controls the initial rate of increase in U for an increase in I_{total}

where:

$$M_n = \sum_{i=1}^{n} U_{n-i+1} B_i \tag{13.2}$$

with $n = 1, 2, 3, \ldots N$

Given a baseline concentration C_b (mg/l), a stream concentration per unit area yield ratio α, an annual catchment N-loading per unit area U_n over a number of N years, with B_i calculated from:

$$B_1 = 0, \quad B_n = A(t_n) - A(t_{n-1}), \quad n = 2, 3, \ldots N \tag{13.3}$$

This used a response function, $A(t)$, calculated from the analytical solution to the advection dispersion equation:

$$A(t) = \text{erfc}\left[\left(\frac{P_e t_a}{4t}\right)^{\frac{1}{2}}\left(1 - \frac{t}{t_a}\right)\right] + \exp\left[\frac{P_e}{2}\right]\text{erfc}\left[\left(\frac{P_e t_a}{4t}\right)^{\frac{1}{2}}\left(1 + \frac{t}{t_a}\right)\right] \tag{13.4}$$

where:
P_e is the Péclet number
T_a is the travel time.

It is useful to note that:

$$\lim_{P_e \to \infty} A(t) = H(t - t_a) \tag{13.5}$$

where:
H denotes the Heaviside step function.

13.3.2 Parameter estimation

By simple graphical translation of the stream-nitrate data superimposed on to the annual catchment N-loading data it is apparent that the travel time, t_a is 37 years (see Figure 13.1). By fixing t_a accordingly, the remaining parameters, P_e, α and C_b are obtained as follows. First a range of P_e values is specified. A value of P_e is selected and a set of M_n values are calculated using Equation (2). Estimates of α and C_b are then obtained by fitting Equation (1) to the observed stream-nitrate concentration data using linear regression. Finally a value of Mean Absolute Error (MAE), between Equation (13.1) with the fitted parameters and the corresponding observed data, is calculated and stored. We deliberately chose the MAE as the objective function as it is not prone to disproportionate influence from individual points compared with, for example, objective functions based on minimising the residual sum of squares such as the Nash-Sutcliffe Efficiency (NSE) or Root Mean Square Error (RMSE).

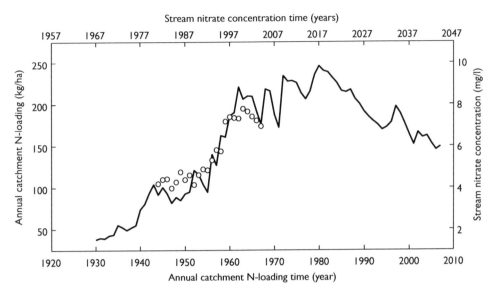

Figure 13.1 Average annual nitrate concentration in the River Piddle at Alton Pancras (1981 to 2004: axes top and right), and estimated catchment nitrogen loading (1930 to 2007: axes left and bottom). Note the 37-year time lag between input (catchment N-load) and output (fluvial concentration rise).

13.3.3   Model parameters

Figure 13.2 shows plots of P_e against corresponding values of MAE, α and C_b: MAE reaches a minimum at $P_e = 1\,418$ and, beyond this, rises to a stable plateau of 0.26 mg/l. Estimates of α and C_b converge with increasing P_e. Essentially it can be said that dispersion (associated with P_e) is negligible. Figure 13.3 shows a comparison of simulated and observed stream-nitrate concentrations using both $P_e = 1\,418$ and $P_e \to \infty$ (a simple delay time model, recall Equation (13.5). Visually it is difficult to choose between $P_e = 1\,418$ and $P_e \to \infty$ as being better or worse. However, $P_e \to \infty$ is arguably preferable owing to its simplicity. The confidence limits shown on Figure 13.3 are based on ±2 times the standard deviation obtained from the linear regression of $M_n\,(P_e \to \infty)$ with the observed stream-concentration data.

It is noticeable that our parameterisation suggests the case $P_e \to \infty$ as a credible parameterisation for the model. Recalling Equation (5), this infers a linear relationship between catchment N-loading and fluvial nitrate-concentration response, lagged by some 37 years. It is, perhaps, surprising given the perceived complexity of long-term catchment-scale nitrate transport processes: non-linear leaching from soils; large residence times for attenuation; and dual porosity aquifer effects. It is generally acknowledged in the literature that a linear model of inputs and outputs is broadly inappropriate, due to the proven non-linearity in observations of nitrate leaching from agricultural soils. The model applied here estimates a net catchment N-loading, due to the inclusion of the hyperbolic function representing crop uptake, such that the assumption of linearity is reasonable. It is also possible, however, that advection is sufficient to explain the link between mean loads and concentrations for an annual time-step.

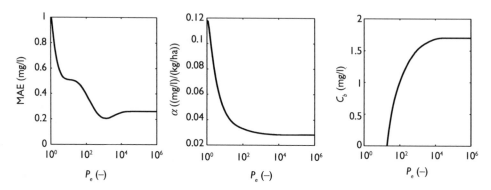

Figure 13.2 Variation of MAE of model fit with Péclet number, and stabilisation of α and C_b parameters as Péclet number increases.

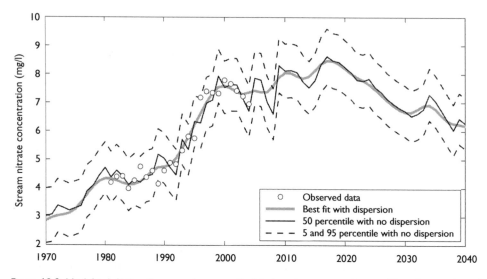

Figure 13.3 Model prediction for annual average fluvial nitrate concentrations at Alton Pancras based on the historical land-use data 1930 to 2008.

13.3.4 Alternative catchment N-loading scenarios

Catchment N-loading calculations were repeated for 1930 to 2007, assuming reduced fertiliser inputs by 25%, 50%, 75% and 100%, given the ubiquity of input controls as a means of addressing diffuse nitrate pollution.

The model output for alternative loading scenarios is shown in Figure 13.4. It is clear that, had fertiliser use been reduced, past, present and future stream-nitrate concentrations would be lower. But it is notable that, even with no fertiliser inputs, rises in stream-nitrate concentration would have occurred between 1960 and 2000. Given the calculated lag time, such rises correspond to changes in catchment N-loading some 37 years earlier.

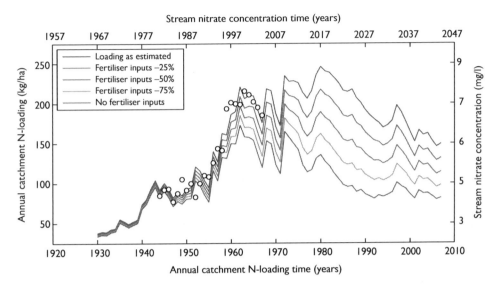

Figure 13.4 Illustration of the effects of changing historical fertiliser-application rates on past, present and future nitrate concentrations in the Piddle at Alton Pancras.

13.4   CONCLUSIONS

We reach four conclusions from this research:

1   In the <10 km² groundwater-dominated watershed considered, a simple comparison of estimated inputs (catchment N-loading) and observed outputs (fluvial concentration response) the time delay between cause and effect is almost four decades.

2   Parameterisation of an (annual time-step) impulse-response model suggests, at this scale, the transport mechanism is dominated by advection, as the best model fits suggest $P_e > 1\ 000$.

3   Taking the case where $P_e \to \infty$ allows the model to be simplified to a linear transformation between inputs and outputs, lagged by 37 years, and model predictions can be specified within 95% confidence intervals. Hence, a prediction of annual average nitrate concentrations up to 2044 may be made from historical land-use and land-management data, which shows that concentrations will continue to rise until almost 2020.

4   Alternative scenarios of historical land use and land management, specifically in respect of fertiliser application rates, suggest that fluvial concentrations between the late 1960s and the early 1980s would still have risen and that fertiliser reductions would not have affected the status quo to a high degree. However, it is clear that post-1990 river concentrations would not continue to rise until 2020 had less fertiliser been used in the past: the full effect of peak-fertiliser usage (1980) will not be seen in the river until around 2017.

The wider relevance of these findings is that nitrate concentrations in groundwater-dominated rivers, especially those fed by the Chalk catchment, may rise for several decades to come due to past activities; any interventions to try and reduce inputs now may not have an effect for many years or even decades. In larger Chalk catchments it has been suggested that the time lag between loading and in-stream response is longer: for example, the time lag in the River Lambourn (261 km^2) is estimated to be between 60 years and 80 years, so we may expect to see a response to on-farm measures first in smaller, headwater Chalk catchments (Jackson *et al.*, 2007; Wade *et al.*, 2008).

REFERENCES

Addiscott, T.M., Whitmore, A.P. & Powlson, D.S. (1991) *Farming, Fertilizers and the Nitrate Problem*. Wallingford, UK, C.A.B. International.

Betton, C., Webb, B.W. & Walling, D.E. (1991) Recent trends in NO_3-N concentrations and loads in British rivers. *International Association of Hydrological Sciences Publication*, 203, 169–180.

Burt, T.P., Arkell, B.P., Trudgill, S.T. & Walling, D.E. (1988) Stream nitrate levels in a small catchment in south west England over a period of 15 years. *Hydrological Processes*, 2, 267–284.

Burt, T.P., Heathwaite, A.L. & Trudgill, S.T. (1993) *Nitrate: Processes, Patterns and Management*. Chichester, UK, Wiley.

Burt, T.P., Howden, N.J.K., Worrall, F. & Whelan, M.J. (2010) Long-term monitoring of river water nitrate: How much data do we need? *Journal of Environmental Monitoring*. [Online] Available from: doi:10.1039/b913003a [Accessed August 2011].

Burt, T.P. & Pinay, G. (2005) Linking hydrology and biogeochemistry in complex landscapes. *Progress in Physical Geography* 29, 297–316.

Casey, H. & Clarke, R.T. (1979) Statistical analysis of nitrate concentrations from the River Frome (Dorset) for the period 1965–76. *Freshwater Biology* 9, 91–97.

Foster, S.S.D. & Crease, R.I. (1974) Nitrate pollution of Chalk groundwater in East Yorkshire – A hydrogeological appraisal. *Journal of the Institution of Water Engineers* 28, 178–194.

Galloway, J.N., Dentener, F.J., Capone, D.G., Boyer, E.W., Howarth, R.W., Seitzinger, S.P., Asner, G.P., Cleveland, C.C., Green, P.A., Holland, E.A., Karl, D.M., Michaels, A.E., Porter, J.H., Townsend, A.E. & Vörösmarty, C.J. (2004) Nitrogen cycles: Past, present and future. *Biogeochemistry* 70, 153–226.

Green, P.A., Vörösmarty, C.J., Meybeck, M., Galloway, J.N., Peterson, B.J. & Boyer, E.W. (2004) Pre-industrial and contemporary fluxes of nitrogen through rivers: a global assessment based on typology. *Biogeochemistry* 68, 71–105.

Heathwaite, A.L., Quinn, P.F. & Hewett, C.J.M. (2005) Modelling and managing critical source areas of diffuse pollution from agricultural land using flow connectivity simulation. *Journal of Hydrology* 304, 446–461.

Howden, N.J.K. (2004) Hydrogeological controls on surface/groundwater interactions in a lowland, permeable Chalk catchment: Implications for water chemistry and numerical modelling. PhD Thesis, Imperial College of Science, University of London, London.

Howden, N.J.K. & Burt, T.P. (2008) Temporal and spatial analysis of nitrate concentrations from the Frome and Piddle catchments in Dorset (UK) for water years 1978 to 2007: Evidence for nitrate breakthrough? *Science of the Total Environment* 407, 507–526. [Online] Available from: doi:10.1016/j.scitotenv.2008.08.042 [Accessed August 2011].

Howden, N.J.K. & Burt, T.P. (2009) Statistical analysis of nitrate concentrations from the Rivers Frome and Piddle (Dorset, UK) for the period 1965–2007. *Ecohydrology*. [Online] Available from: doi:10.1002/eco.39 [Accessed August 2011].

Howden, N.J.K., Neal, C., Wheater, H.S. & Kirk, S. (2010) Water quality of lowland, permeable Chalk rivers: The Frome and Piddle catchments, west Dorset, UK. *Hydrology Research* 41, 75–91.

Jackson, B.M., Wheater, H.S., Wade, A.J., Butterfield, D., Mathias, S.A., Ireson, A.M., Butler, A.P., McIntyre, N.R. & Whitehead, P.G. (2007) Catchment–scale modelling of flow and nutrient transport in the Chalk unsaturated zone. *Ecological Modelling* 209, 41–52.

Jackson, B.M., Browne, C.A., Butler, A.P., Peach, D., Wade, A.J. & Wheater, H.S. (2008) Nitrate transport in Chalk catchments: Monitoring, modelling and policy implications. *Environmental Science and Policy* 11, 125–135.

Mathias, S.A., Butler, A.P., McIntyre, N. & Wheater, H.S. (2005) The significance of flow in the matrix of the Chalk unsaturated zone. *Journal of Hydrology* 310, 62–77.

Mathias, S.A., Butler, A.P., Ireson, A.M., Jackson, B.M., McIntyre, N. & Wheater, H.S. (2007) Recent advances in modelling nitrate transport in the Chalk unsaturated zone. *Quarterly Journal of Engineering Geology and Hydrogeology* 40, 353–359. November 2007, Geological Society, London, UK.

Meybeck, M. (1982) Carbon, nitrogen and phosphorus transport by world rivers. *American Journal of Science* 282, 401–450.

Roberts, G. & Marsh, T. (1987) The effects of agricultural practices on the nitrate concentrations in the surface water domestic supply sources of Western Europe. *Journal of the International Association of Hydrological Sciences* 164, 365–380.

Royal Society (1983) *The Nitrogen Cycle: A Royal Society Discussion Held on 17 and 18 June 1981*. London, UK, Royal Society.

Wade, A.J., Butterfield, D. & Whitehead, P.G. (2006) Towards an improved understanding of the nitrate dynamics in lowland, permeable river systems: Applications of INCA-N. *Journal of Hydrology* 330, 185–203.

Wade, A.J., Jackson, B.M. & Butterfield, D. (2008) Over-parameterised, uncertain 'mathematical marionettes' – How can we best use catchment water quality models? An example of an 80-year catchment-scale nutrient balance. *Science of the Total Environment* 400, 52–74. [Online] Available from: doi:10.1016/j.scitotenv.2008.04.030 [Accessed August 2011].

Weatherhead, E.K. & Howden, N.J.K. (2009) The relationship between land use and surface water resources in the UK. *Land Use Policy* 26, S243–S250.

Whitehead, P.G., Wilby, R.L., Butterfield, D. & Wade, A.J. (2006) Impacts of climate change on nitrogen in a lowland Chalk stream: An appraisal of adaptation strategies. *Science of the Total Environment* 365, 260–273.

Whitmore, A.P., Bradbury, N.J. & Johnson, P.A. (1992) Potential contribution of ploughed grassland to nitrate leaching. *Agriculture, Ecosystems and Environment* 39, 221–233.

Chapter 14

Hydrogeological study for sustainable water-resource exploitation – Ibo Island, Mozambique

Ester Vilanova, Eva Docampo, Cristina Mecerreyes &
Jorge Molinero
Amphos 21 Consulting SL. Ps Garcia i Fària, Barcelona, Spain

ABSTRACT

Ibo Island, located in northern Mozambique, is a coral atoll, less than 10 km in length and nowadays featuring a growing tourism industry and business market. A hydrogeological study has been carried out with the aim of providing tools for sustainable water-resource management. The existing aquifer is recharged from rainwater infiltration and characterised by a double porosity through the main unaltered rock and well-developed karst channels, with the associated high risk of water pollution. As a general pattern, groundwater flows from the centre of the island to the coastline, through submarine discharges with increasing seawater intrusion and bacteriological contamination. The freshwater/seawater interface is characterised by a mixing zone depending on annual seasonality but also on tide periodicity. Renewable fresh groundwater resources were quantified using an aggregated water-balance model for the island and calibrated to obtain an annual aquifer recharge of 146 mm/yr. Freshwater usable thickness was estimated to be up to 5 m concluding that it is feasible to extract up to 0.6 Hm^3/yr taking into account the extreme vulnerability of the Ibo Island aquifer to prolonged drought episodes. Final recommendations for sustainable development were elaborated to institutions at different government levels responsible for implementing the sustainable development goals.

14.1 INTRODUCTION

Mozambique has achieved significant improvement in water access, increasing its rural water-supply coverage from 30% up to 60% in 10 years (DNA, 2011). Water-resource assessment and integrated management has recently been included in the national agenda to ensure sustainability and efficiency. This hydrogeological study of Ibo Island is a case study that can be replicated in rural areas of countries with low data availability and water scarcity or resource vulnerability as well as with saline-intrusion problems.

14.1.1 Context

Ibo Island is part of the Quirimbas Archipelago, a chain formed by around 28 islands aligned north-south, covering 400 km alongside the littoral region of Cabo Delgado province in Mozambique (Figure 14.1). At its nearest point the island is 375 m away from the mainland The surface area of the island is about 10 km² (3.6 km × 4.5 km) and it is part of the Quirimbas National Park. The topography of Ibo Island is mainly flat, with the highest elevation point being located at San Antonio fort (around 10 m amsl) according to the urban plan of the island. Ibo village (the formal urbanised area) is located in the south-western edge of the island and it is divided into three neighbourhoods, namely Cimento, Cumuamba and Rituto. Quirambo neighbourhood, situated on another island, also belongs to Ibo village (Figure 14.1). The rest of the island is without planned urbanised areas and is referred to as *informal* settlement.

In 2007 provincial and district institutions together with the technical assistance of the Spanish Agency for International Cooperation and Development (AECID) developed the Urban Planning and the Detail Urban Plan (GPCB, 2007 and 2008) for Ibo village, providing key guidelines to guide, manage and control the future development in the village. Within this process, water and sanitation issues were raised as a main factor to ensure future and sustainable human and economic development on the island. The consultancy firm Amphos 21 was requested by the Spanish AECID to undertake the quantitative and qualitative water-resource assessment for the island, its management and regulatory recommendations and finally the design of the water-supply system. From June 2009 to June 2010, a basic hydrogeological study with the development of water-management scenarios and final recommendations was carried out, including field work, analysis of institutional capacity and a monitoring network set-up. Several presentations of the results to local and national authorities were carried out during 2010 and 2011 in order to raise awareness to the requirement for water control and supply regulation for sustainable development on Ibo Island. The population of Ibo Island was around 3 650 in 2009, and is expected to increase by 1 000 people over 5 years and by more than 5 000 people in 15 years, in terms of a regional average population growth rate of 2.5%. Presently, hotel beds number around 130, distributed over five establishments. Conservative estimates foresee the provision of up

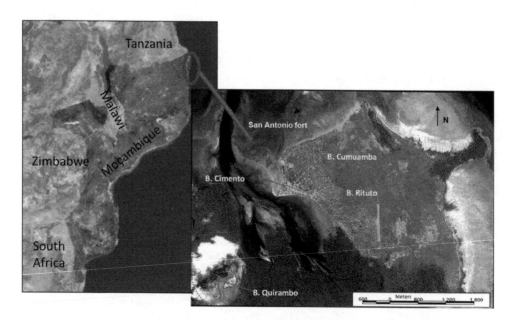

Figure 14.1 Location of Ibo Island and study zone.

to 70 additional beds in 2015, but field interviews suggest that the number could be higher, if tourism sector expectations are met.

The water supply for Ibo village and informal areas is provided by manually-excavated public or private traditional wells, ranging from 2 m to 8 m in depth, and water is extracted manually (hand pumps, buckets) (Figure 14.2 and Figure 14.3). Only 4% of the wells are equipped with electrical or solar pumps, mainly those supplying water to hotels. A power supply connection to the mainland electricity grid is promised for the coming years, to change the conditions of the island and assist its development. In this context, the rational planning and exploitation of Ibo Island's water resources are essential to achieve sustainable development, whereas extremely uncontrolled exploitation would certainly lead to a decline in water reserves, together with pollution of wells due to seawater intrusion (coastal aquifer overexploitation) and human activities (sanitation, animals).

14.1.2 Objectives

The main aim of this work was to conduct a comprehensive and rigorous hydrogeological study in order to guarantee the sustainable water exploitation and management on Ibo Island, together with institutional capacity-building and appropriate monitoring tools set up during implementation. The final expected result was to deliver a sound water-resource management plan and to make policy recommendations to institutions responsible for implementation of the policy.

Figure 14.2 Open, manually excavated well.

Figure 14.3 Protected well.

14.2 METHODOLOGY

Very limited previous and current data were available for the area, mainly comprising isolated references to water availability and lacking qualitative data analysis. Water-resource assessment in this study was developed from compilation and interpretation of hydrometeorological data, field work and interpretation of compiled data. The field campaigns included geological characterisation, an inventory of water points, water sampling, in *situ* water analysis and pumping tests, water-table monitoring, and water-quality determination. As a result, geological, potentiometric and hydrochemical studies were conducted, maps were drawn up, and a water budget of the island was estimated together with available water resources.

In order to obtain the necessary data to support the present work, two field campaigns were carried out. The first campaign was conducted during the dry season between 23 August 2009 and 9 September 2009. The second took place in the period between 25 January 2010 and 5 April 2010. All existing wells were visited and fully characterised using an identification sheet for each water point, adding the official water point mapping sheet at national level hydrogeological missing data (Figure 14.4). The water-well inventory was updated from 152 (DNA, 2009) to 231 records and were all plotted on a specific water point map. Moreover, more knowledge was gained relating to water-service provision in place and all the well characteristics were recorded. Besides UTM coordinates and physical data (diameter, depth), potentiometric levels and physicochemical parameters were measured *in situ*, thus allowing future interpretation and monitoring. Ten water samples were collected for complete hydrochemical analysis in the Environmental, Hygiene and Medical Examination Centre (CHAEM) laboratory in Pemba (Mozambique) and Manresa Technologic Centre (CTM) in Barcelona (Spain). In order to allow water-resource quantification and quality characterisation, a monitoring network including 26 wells was established on the island. This network was managed for almost one year by local authorities (previously trained by Amphos 21) and equipped with a portable water-level sensor and an *in-situ* water-quality probe to measure temperature, pH and electrical conductivity. Also, a 3-day technical seminar on basic hydrogeology and water resources management was held at provincial level with representatives from the province, districts, basin organisation and local NGO staff, funded by AECID in collaboration with Amphos 21.

A Geographical Information System (GIS) was implemented using gvSIG 1.9., open software initially developed by a regional Spanish government authority (Generalitat Valenciana) for institutional use, and nowadays widely used in places such as Portugal or Latin American countries. The approach was to introduce technologies easily transferable to local technicians and institutions, in order to promote appropriation and development of their capacities. Other specific hydrogeological

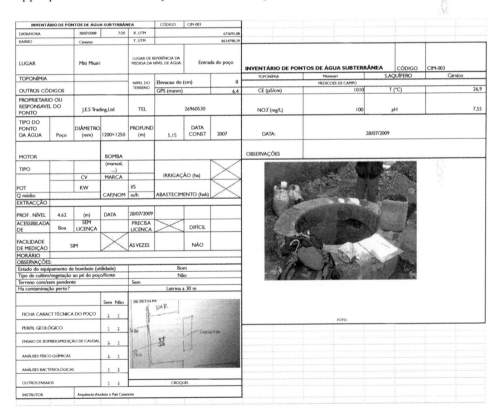

Figure 14.4 Water point identification data sheet.

software used during the study was Visual Balan 2.0 (Samper *et al.*, 2005) for water budget and data calibration or Ephebo – MARIAJ_IV (Carbonell *et al.*, 1997) to support hydraulic pumping test interpretation.

14.3 HYDROGEOLOGICAL STUDY

In terms of the UNESCO classification (UNESCO, 1991) Ibo Island belongs to the *small islands* group since its width is not greater than 10 km. The importance of this classification comes from the fact that small islands have distinct hydrological and water-resource assessment, developmental and management problems, unlike those of medium and large islands.

14.3.1 Geology

Ibo Island is part of the sedimentary basin of Rovuma-Mozambique, and at the same time integrated within the East-African peri-continental basin. Regionally, Ibo Island is linked to a tidal plain connected to river estuaries in the zone and surface coastal channels separating the island from the continent, creating a mangrove-dominated area. Constituent materials are basically carbonate deposits (coral reef), originating from organic material and quaternary shell fragments of Pleistocene to Holocene age. These deposits show massive/irregular stratification and cover 70% of the surface area of the island. They are overlain by terrigenous siliciclastic quaternary sediments (sand, silt and clay carbonate) (Figure 14.5) that cover approximately 20% of the island and originate from the tidal plain. A small geophysical study carried out by the National Water Directorate (DNA) indicated point values of around 5 m for the upper clastic levels and between 15 m and 20 m for the deeper carbonate layer in the urban area of the island. Nevertheless, this was a localised study and more precision is required to infer the thickness of each layer. The karst development on the island is high, with elevated numbers of fractures and typical dissolution karst structures in the coral rock basement. This factor is of utmost importance, since it influences groundwater-resource distribution throughout the island, as well as its annual availability. Because of its climate regime, geological characteristics and plain topography, surface-water resources are rarely found on the island.

14.3.2 Groundwater

Karstic carbonate aquifers are extremely heterogeneous with a distribution of permeability that spans many orders of magnitude. They often contain open conduit flow paths that allow the rapid transport of groundwater. The conduit system interconnects with the groundwater stored in fractures and in the granular permeability of the bedrock, creating a so-called double porosity material. On islands of carbonate coral rocks, caves and caverns form as a result of karstic dissolution of carbonate rocks, and groundwater creates freshwater lenses overlying seawater (Wheatcraft and Buddemeier, 1981). These freshwater discharges below sea level at the coastal areas are driven down by precipitation which infiltrates the centre of the island, as shown in Figure 14.6. The depth of the freshwater lenses on the island is not clearly defined as there is vertical mixing with underlying saline water, creating the so-called transition

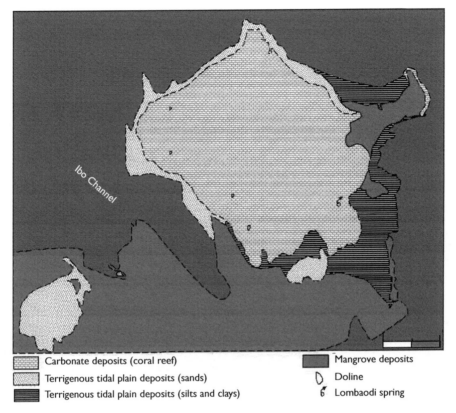

Carbonate deposits (coral reef)

Terrigenous tidal plain deposits (sands)

Terrigenous tidal plain deposits (silts and clays)

Mangrove deposits

Doline

Lombaodi spring

Figure 14.5 Geology of Ibo Island.

zone (Figure 14.7). Transition zone thickness may be double that of the freshwater lens, but may then disappear under extreme conditions (drought, overexploitation). For this study, an electrical conductivity of 2 000 μS/cm is assumed as the upper value for human consumption, as it is also the established limit set by Mozambican regulations on water quality (Diploma Ministerial No. 180).

Potentiometric surface maps for dry and rainy seasons (hydrological year 2009–2010) were built, using weekly water-level measurements collected during field campaigns in the monitoring network (26 representative wells previously selected) (Figure 14.7). Wells depth ranged from 1 m to 10 m, the later being the deepest ones in the southern part of the island, in Rituto neighbourhood. Water-table elevations varied from negative values in certain areas to around 7 m amsl. In RIT-046, the unique well located in the centre of the island, the potentiometric level remains nearby 4 m amsl throughout the hydrological year, showing only 0.5 m variation between seasons and slight tidal effects. Tidal effects were evaluated in some wells, showing influences lower than 5 cm. Seasonal level variations ranged from 10 cm to 80 cm, with 80% of the monitoring network points being below 50 cm annual variation (hydrological year 2009–2010). As a general pattern, groundwater flows from the centre of the island to the coastal line, through submarine discharges. The potentiometric levels follow the surface topography

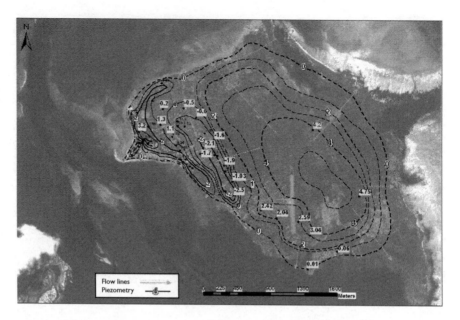

Figure 14.6 Potentiometric map (dry season 2010).

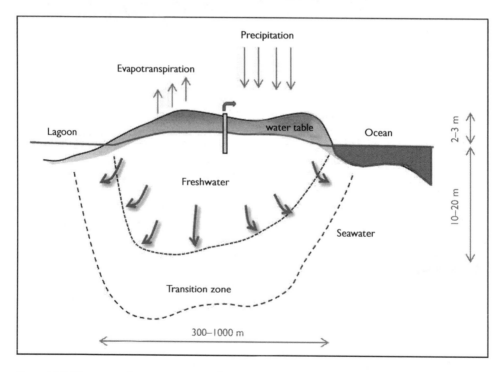

Figure 14.7 Diagrammatic representation of a vertically coupled model (Wheatcraft and Buddemeier, 1981) used to describe freshwater lens behaviour and mixing with underlying saline water on small coral islands.

except in some areas of preferential recharge or intensive groundwater exploitation. The recharge takes place over the entire surface area of island, along the main temporal water courses and preferential recharge areas as in the western zone, where a water-recharge mound has been identified. Near this area, and coincident with the hotel's location, a water-level depression was observed, clearly identified as a pumping cone, with a 0.5 m reduction compared with natural water-table elevations.

The Ibo Island wells can be classified in terms of their water quality and water-abstraction methods and therefore their water-supply reliability as follows: open/protected; active/dry/abandoned; and equipped with bucket/hand-pump/solar or electrical pump. It was established that 55% of the wells are open, active and not equipped with a pump while only 16% of them are protected with a cover and hand pump, and 4% of the wells are equipped with a submersible pump. Only one natural spring was identified, located in the south-eastern part of the island. Some water-quality parameters were analysed including bacteriological pollution (number of coli forms) and nitrate content in water (mg/l), revealing important levels of water contamination, with a strong link to well-construction characteristics and utilisation aspects (Figure 14.8). Seasonal effects were also confirmed: During the rainy season, freshwater rapidly infiltrates the ground and dilutes pollution in some highly contaminated wells resulting in less polluted water. Moreover this sub-surface flow also transports contaminants to non-polluted wells, even protected ones increasing their pollution concentration (Figure 14.9). This phenomenon is especially risky when unlined pit latrines are located upstream of groundwater flow. Nitrate contamination (above 50 mg/l) is mainly found

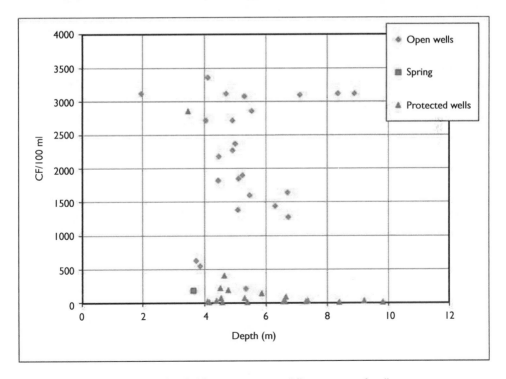

Figure 14.8 Coliforms in water in different types of wells.

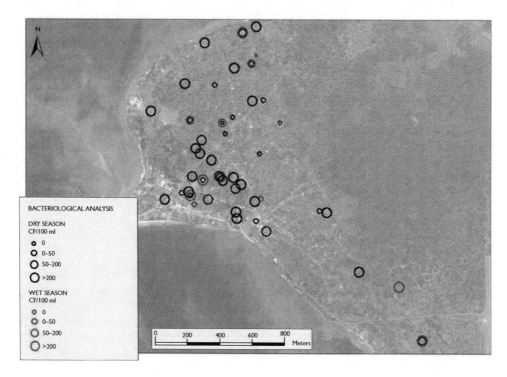

Figure 14.9 Bacteriological contamination in dry and rainy season.

in informal areas, probably related to cattle. After rains, 25% of the wells show concentrations of nitrate higher than 50 mg/l with a maximum of 350 mg/l.

Isovalue maps of electrical conductivity were drawn using measurements in wells as an indicator of aquifer salinisation. These maps show saline intrusion plumes in urban zones and the western part of the island that can be related to geological preferential flow paths (porosity, karstic dissolution channels) together with groundwater exploitation near the coast (where water demand is higher). In general, saline water is observed in tidal marshes and mangrove areas. Increased salinisation of groundwater occurs mainly during the dry season (Figures 14.10 and 14.11), clearly indicating the high vulnerability of the aquifer to the reduction in recharge (precipitation).

14.3.3 Aquifer recharge and water budget

A water-budget calculation is a prerequisite to estimating the aquifer recharge on the island, and thus to quantifying water-resource availability and delivering policy recommendations for the sustainable exploitation of groundwater. Ibo Island recharge is derived totally from precipitation. To calculate the total recharge to groundwater, hydrometeorological data, vegetation and soil characteristics of the island were introduced into the software Visual Balan 2.0 (Samper *et al.*, 2005). This is a semi-distributed hydrological model for water-resource assessment, developed by the Hydrology Group of the University of A Coruña (Spain). Visual Balan solves

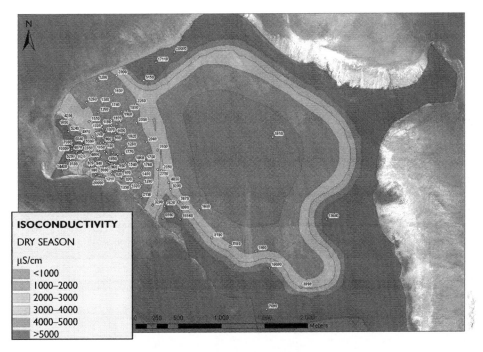

Figure 14.10 Map of electrical conductivity distribution for dry season (2009).

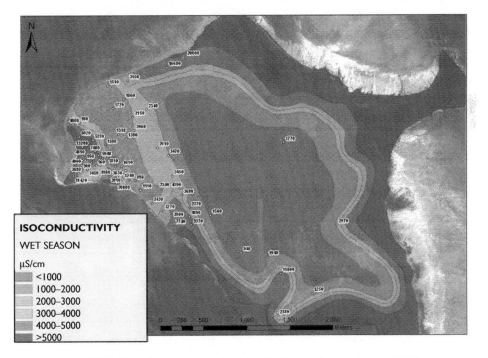

Figure 14.11 Map of electrical conductivity distribution for wet season (2010).

the water balance in the soil, the unsaturated zone and the aquifer calculating the components in a sequential manner. The calibration task was performed using three water points monitored during this project, regarded as being representative of the aquifer behaviour. Soil and vegetation parameters were estimated and adjusted from in-field observations, bibliography data and model requirements. Hydrometeorological data were compiled for 22 years (1988–2010), using several information sources (Pemba, Ibo and Quirimba) and establishing correlations to compare and complete the series. The climate of Ibo Island is humid and subtropical, with an average temperature of 26°C. 86% of the average annual precipitation is received during the rainy season (between November and April) with temperatures ranging from 24°C to 28°C. Field data from control points during the 2009 rainy season showed that the aquifer response to precipitation episodes is rapid and uniformly distributed. Island vegetation is thick and abundant, formed by coastal bushes, palm trees and baobabs, with significant rain-interception capacity. Soil types are classified under Eutric Leptosols and Calcaric Arenosols/Regosols (SMU-FAO CD09-Lpe/Arc/RGc) with slopes less than 5%, in terms of the Harmonized World Soil Database. In general, soils on the island are shallow and undeveloped (average 40 cm), extremely stony (leptosols), or unconsolidated calcareous materials (arenosols), and with low water-retention capacity.

The results of the model, once calibrated for the hydrological year 2009–2010, gave the potential and actual evapotranspiration (PET, AE), vegetal interception, surface runoff and subsurface flux quantifications, using an annual average precipitation of 1 028 mm (Figure 14.12). The model satisfactorily reproduced the measured levels in the three wells (Figure 14.13) when calibrated by adjusting the storage coefficient, the discharge time of the aquifer and the parameters for the edaphic and unsaturated soils fractions. The calibrated model was then applied to the 1988–2010 period (annual average precipitation of 904 mm), obtaining an estimated annual aquifer recharge of 146 mm/yr after considering all parameter uncertainties. The volume of groundwater withdrawals was estimated at 0.03 Hm³/yr. Different information sources were taken into account to calculate this value. An average of 22 l/person · d was used to calculate annual extractions from traditional wells (public or private) based on data from field observations, interviews and also knowledge gleaned from experts. Moreover, consumption figures from hotels, agriculture and cattle were estimated and added to obtain final calculations on groundwater extractions. To study the future water demands and to evaluate the future sustainability of the natural system, the 2015 development scenario was considered. This scenario includes an annual population growth of 2.5% and an additional increase in water consumption per person and economic activities (tourism), resulting 0.23 Hm³/yr as total expected groundwater extractions.

14.3.4 Annual groundwater reserves

As has been previously detailed, groundwater in coral atoll islands creates freshwater lenses overlying seawater and generating a vertical mixing transition zone (White and Falkland, 2010). Maximum freshwater lenses and transition zone thickness (H_u, δ_u, respectively) for Ibo island centre were estimated using Equations (14.1) and (14.2) (Volker et al., 1985; White and Fackland, 2010; Wooding, 1984):

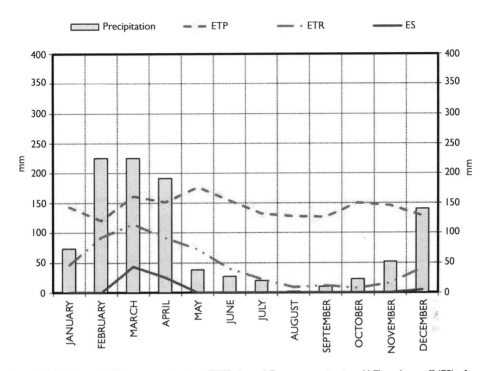

Figure 14.12 Potential Evapotranspiration (PET), Actual Evapotranspiration (AE) and runoff (ES) after calibration.

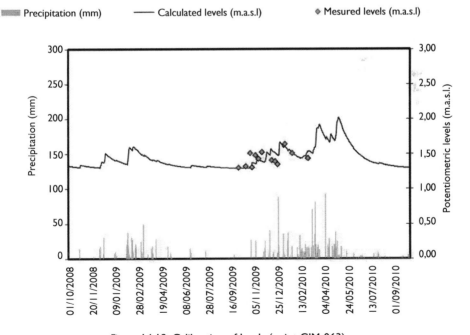

Figure 14.13 Calibration of levels (point CIM-063).

$$H_u = \frac{W}{2} \cdot \sqrt{(1+\alpha) \cdot \frac{R}{2 \cdot K_0}}$$ (14.1)

$$\frac{\delta_u}{H_u} = \frac{K_0}{R} \sqrt{\left(\frac{D}{\alpha \cdot W \cdot K_0} \right)}$$ (14.2)

where:

W is the island width (2.5 km)

K_0 is the uniform horizontal hydraulic conductivity (6 m/d to 13 m/d, based on field-pumping tests)

α is the $(\rho_s - \rho_0)/\rho_0$ coefficient (1.3)

D is the dispersion coefficient ($2.84 \cdot 10^{-4}$ m²/yr)

Using 146 mm/yr as the annual recharge (R), H_u varies from 5.0 m to 7.3 m (decreasing when K_0 increases). H_u is calculated assuming a stationary sharp freshwater/seawater interface and the H_u depth is about half of the transition zone (assuming a gradual mix). This value should be considered qualitatively and as an indicator of aquifer vulnerability. The transition zone, δ_u, was calculated using Equation (14.2) and assuming a stationary system. The resulting transition zone thickness is between 5 m and 10 m.

As the H_u model assumes a sharp interface, a reduced freshwater usable thickness (H_{wu}) was calculated for this project, thus preventing the extraction of water from the deeper transition zone. These calculations were performed for several aquifer-recharge scenarios and considering different annual precipitation values obtained from available data series (1988–2010), as well as using two different K_0 values for the aquifer. The K values were calculated by conducting various pumping tests on the island.

The usable freshwater thickness (H_{wu}) is the groundwater resource available for water supply on the island, ranging from 2 m to 5 m during the period analysed. If a 50% drought episode occurs, H_{wu} may decline to less than 1 m. This fact shows the extreme vulnerability of the Ibo Island aquifer to prolonged drought episodes. Finally, the freshwater response to uniform distributed pumping was analysed, calculating H_p (usable water under $q = (Q/A)/R$ pumping) using Equation (14.4)) and concluding that it is feasible to extract up to 0.6 Hm³/yr (three times the development scenario estimations) if a shallow well field is implemented over a 4 km² area located in the centre of the island, with low pumping rates and carefully monitoring water levels and electrical conductivity.

$$H_{wu} = H_u - \frac{\delta_u}{2}$$ (14.3)

$$H_p = \frac{W \cdot \sqrt{(1-q)}}{2} \cdot \sqrt{(1+\alpha) \cdot \frac{R}{2 \cdot K_0}} = \sqrt{(1-q)} \cdot H_u$$ (14.4)

It should be noted that the development of underground karst structures can affect the pumping scenario and other pumping areas will have to be considered if important karstic channels are detected in the prospection zone.

14.3.5   Conclusions and policy recommendations

Final conclusions and recommendations for water-resource management and protection on the island were presented to institutions, residents and stakeholders in the province:

- Significant levels of faecal and nitrate pollution of the groundwater in the village were detected. This is mainly due to the elevated number of unprotected or abandoned wells in the village, allowing contaminants access from the surface, in a geological and karstic context that also facilitates rapid subsurface water circulation, originating for example from nearby pit latrines. Seasonal variability also confirms this mechanism.
- Groundwater sources available for pumping occur in freshwater lenses overlying seawater, with an approximate usable thickness of 2 m and a seasonal maximum of 5 m. This thickness is highly vulnerable to drought episodes where vertical mixing in the transition zone greatly increases, and to local pumping leading to saline intrusion.
- Annual recharge is totally dependent on precipitation episodes and subsequent infiltration into the aquifer, being greater in the centre of the island and in the absence of vegetation and developed soils.
- Groundwater extraction to provide the island with the water required in the developed scenario for 2015 would be feasible if a water supply was established from several shallow wells distributed over a 4 km^2 area in the centre of the island and with low pumping rates.

Policy recommendations:

- To establish a water-supply system from shallow wells in the centre of the island, together with a monitoring system to continuously measure levels and electrical conductivity in the wells. As local karst structures can affect the wells and aquifer behaviour, an exploratory study of underground structures (cavities and channels) is strongly recommended.
- To implement an improvement plan for water points, select the wells that are in good condition and strategically located to be protected and integrated in the water-supply system, while cleaning and plugging the abandoned or polluted wells.
- To construct at least three piezometers in the centre of the island to execute pumping test and observations prior to the establishment of water-supply wells field.
- To reinforce local regulations to avoid the extraction of aggregates in the centre of the island.
- To establish a protected area around the wells located in the centre, where potential contaminating activities should be totally forbidden.
- To characterise potential pollution resulting from poor sanitation threatening the aquifer and to implement appropriate technologies to prevent the contamination of wells, together with protection perimeters.
- To implement a strong information and sensitisation campaign focused on population awareness of water-resource protection and management, fostering participation, and promoting the improvement of hygiene behaviour, and environmental education.

- To establish a public-private management structure for water-system exploitation, including the participation of hotels and water-user committees in the neighbourhoods.
- To measure delivered water and clearly establish water volumes allowed per household, economic activities and other uses, differencing water fees and taxes.
- To include a pro-poor approach from the design phase, with public standpoints and various subsidies depending on economic capacity.

As result of the present study, a new project to design the water-supply system was approved and funded by the Spanish AECID for 2011, aimed at integrating all the recommendations and produce the final documents that will lead to water-supply system procurement and execution.

ACKNOWLEDGEMENTS

Patricia Casanova and Isequiel Alcolete for field work and dedication, and Spanish Agency for International Cooperation and Development (AECID), particularly to Ph.D. Jesus Pérez.

REFERENCES

Carbonell, J.A., Pérez-Paricio, A. & Carrera, J. (1997) MARIAJ_IV, Code for pumping tests automatic calibration. User's guide (draft) E.T.S.I. Caminos Canales y puertos. Barcelona, Spain, Universitat Politècnica de Catalunya.

Diploma Ministerial No. 180 (2004) de 15 de Setembro de 2004. Regulamento sobre a Qualidade da Água para o Consumo Humano (Regulation on water quality for human consumption).

DNA, Direcção Nacional de Águas (2009) Ibo island water study. Internal report.

DNA, Direcção Nacional de Águas (2011) Relatório Anual de Avaliação do Desempenho do Sector de Águas 2009–2010 [Annual report of performance evaluation of water sector]. Internal report.

GPCB (Governo da província de Cabo Delgado) (2007) Plano de Urbanização da vila do Ibo (Urban plan for Ibo village).

GPCB (Governo da província de Cabo Delgado) (2008) Plano de Pormenor da vila do Ibo (Detail urban plan for Ibo village).

Samper, J., Hugget, L.l, Ares, j & García-Vera, M.A. (2005) Manual del usuário del programa Visual Balan 2.0 (User's manual). Madrid, Spain, ENRESA. pp. 139.

UNESCO (1991) Hydrology and Water Resources of Small Islands: A Practical Guide. Paris, France, UNESCO.

Volker, R.E., Mariño, M.A. & Rolston, D.E. (1985) Transition zone width in ground water on ocean atolls. Journal Hydraulic Engineering 111, 659–676.

Wheatcraft, S.W. & Buddemeier, R.W. (1981) Atoll island hydrology. Ground Water 19, 311–320.

White, I. & Falkland, T. (2010) Management of freshwater lenses on small Pacific islands. Hydrogeology Journal 18, 227–246.

Wooding, R.A. (1964) Mixing-layer flows in a saturated porous media. Journal of Fluid Mechanics 19, 103–112.

Chapter 15

In situ nitrate removal from groundwater using freely available carbon material at an industrially polluted site

S. Israel[1], A. Rosenov[2], G. Tredoux[1] & N. Jovanovic[1]

[1]Council for Scientific and Industrial Research, Natural Resources and Environment, Applied Hydrosciences Group, Stellenbosch, Western Cape, South Africa
[2]Faculty of Agriculture, Soil Science Department, University of Stellenbosch, Stellenbosch, South Africa

ABSTRACT

Groundwater pollution by nitrate is a known problem which occurs worldwide. In its nitrate (NO_3^-) form, nitrogen can infiltrate groundwater and also contributes to surface-water quality problems. Safe methods of treating nitrates with minimal costs are required technological endeavours in poor rural households in Africa. The basic principle for *in situ* denitrification is that a carbon source is placed perpendicular to the groundwater flow such that flow occurs through the carbon source layer. A chemical reaction mediated by naturally available bacteria then accelerates the natural process of denitrification. This paper discusses laboratory experiments and up-scaled field application of the method. A 2.15 m diameter × 1.37 m height tank filled with sawdust was buried to 3 m depth at a site with high industrial-level nitrate concentrations of above 1 000 mg/l as NO_3^-. Boreholes in the vicinity of the carbon source were monitored for a number of parameters bi-weekly. Results showed that 100% of nitrate was removed from the treatment zone within 2 d. It was successfully demonstrated that *in situ* denitrification using sawdust as a cheaply available and slowly degradable carbon source can effect total removal of high industrial-level nitrate concentrations.

15.1 INTRODUCTION

Groundwater pollution by nitrate is a known problem which occurs all around the world. In its nitrate (NO_3^-) and other forms, nitrogen can move through the soil into groundwater. Nitrogen can also contribute to surface-water quality problems (Canter, 1997). Nitrate concentration in groundwater is of concern due to potential effects on human health as well as effects on livestock, crops, and industrial processes at high concentrations (DWAF, 1996). A condition called methaemoglobinaemia also known as 'blue baby syndrome' results from the ingestion of high concentrations of nitrate in its inorganic form (ITRCWG, 2000). Infants, children and adults suffering from maladies or treatments that lower the levels of stomach acid, are vulnerable to methaemoglobinaemia (ITRCWG, 2000). Data from case studies of methaemoglobinaemia (Addiscott and Benjamin, 2004), show that the threat is real and most cases are reported where concentrations of nitrate are above 20 mg/l NO_3^- as N (DWAF, 1998), which is the South African water-quality guideline value for maximum allowable nitrate concentration in drinking water (Tredoux *et al.*, 2001) Above 300 mg/l as N, nitrate poisoning may result in the death of livestock consuming water. At lower concentrations, other adverse effects occur in animals and these include increased incidences of still-born calves, abortions, retained placenta, cystic ovaries, lower milk production, reduced weight gains and vitamin A deficiency. Recommended levels of nitrate for stock watering (livestock and poultry) in the US is below 100 mg/l as N (ITRCWG, 2000; Innovative Technology, 2000).

In situ biological denitrification refers to processes of enhancing the natural system's ability to denitrify water. The method is capable of denitrifying groundwater, wastewater, treated effluent from wastewater works and other polluted water. It requires the addition of a suitable substrate and normally a carbon source if heterotrophic denitrification is desired. Autotrophic denitrification takes place when reduced sulphur compounds, ferrous iron or hydrogen is added to the subsurface (Mateju *et al.*, 1992; Mercado *et al.*, 1988). Permeable reactive barrier walls are constructed by digging a trench or hole of suitable size and configuration perpendicular to the groundwater flow direction, and mixing aquifer material with organic matter, e.g. sawdust, woodchips, which acts as a carbon source to stimulate denitrification (Schipper *et al.*, 2004; Schipper and Vojvodic-Vukovic, 2000; 2001; Robertson and Cherry 1995; Robertson *et al.*, 2000, Robertson *et al.*, 2003; Gavascar, 1999).

The efficient functioning of a reactive barrier can often depend on the proper emplacement of the actual barrier and its effectiveness at allowing flow of source water, be it groundwater or wastewater through it. Many studies have shown successful denitrification all over the world. However, denitrification at some of the sites was not always as successful, and often scientists who have successfully denitrified their source water through a permeable reactive barrier in one location have had failures in other areas using similar techniques. Successful denitrification of groundwater has been practised for more than a decade in New Zealand (Schipper and Vojdovic-Vukovic, 2000; 2001; Shipper *et al.*, 2004). The latest attempted denitrification of non-point sources of nitrate from shallow groundwater failed due to hydraulic constraints on the performance of the denitrification wall (Schipper *et al.*, 2004). Denitrification walls are most successful and effective at protecting downstream water quality when as much groundwater as possible is intercepted by them (Schipper *et al.*, 2004; Barkle *et al.*, 2008; Robertson *et al.*, 2003). Hence the hydraulic properties of both the wall and aquifer are integral properties to be monitored and assessed throughout the life of an operation (Barkle *et al.*, 2008).

Field- and pilot-scale denitrification studies elsewhere in the world show very promising results (Schipper *et al.*, 2004; Schipper and Vojvodic-Vukovic 2000; 2001; Robertson and Cherry, 1995; Robertson *et al.*, 2000), and testing of this technology under field conditions in South Africa is therefore a required progression given the high nitrate risks to drinking water in the country. Our study approach involved firstly a laboratory component in which the most suitable carbon source was identified through controlled experiments. The study was then scaled up to field conditions where the technology was implemented on a pilot scale. The main objectives of the experiment were to test the feasibility of field denitrification at a site with high industrial-level nitrate concentrations in groundwater. Testing whether denitrification will take place successfully where clay is present was another objective as this presented some uncertainty. Key implications of field implementations, limitations that may arise and lessons learnt during the field implementation are discussed.

15.2   SITE CHARACTERISATION: PRE-FIELD IMPLEMENTATION PROCEDURES

Laboratory evaluations took the form of treatability tests done on the bench scale using various carbon sources (Israel *et al.*, 2007) and laboratory flow through testing with a selected carbon source from treatability tests (Israel *et al.*, 2005). The laboratory bench-scale treatability experiment was used to decide on a most suitable carbon source for the desired use. A long-term, low-maintenance and slowly degradable carbon source with effective denitrification was deemed ideal. Carbon sources used during treatability tests included glucose, methanol, maize meal and sawdust. Sawdust was selected as a carbon source for observations in the field and laboratory using standard flow-through experiments due to its slow degradability, effective denitrification and minimal by-products (Israel *et al.*, 2005; Israel, 2007).

The laboratory flow-tank experiment was used to consider design parameters for field-scale testing, e.g. amount of sawdust required based on requirements for denitrification. Barrier thickness or barrier concentration was determined from calculations using fundamental stoichiometric relationships and incorporating flow and area. Nitrite was produced during the experiment which indicated incomplete denitrification. The sawdust denitrification requirement for the flow-tank experiment was not met for the full duration of the experiment. The maximum period of denitrification occurred during the initial 10 d and between days 20 and 40 of the experiment. This could be due to the slow and intermittent release of available carbon for denitrification, the high nitrate concentration used during this experiment and the period of monitoring used.

The site used for field testing was in the Somerset West area of the Western Cape Province (Figure 15.1). Historical reports for the site and available data were consulted

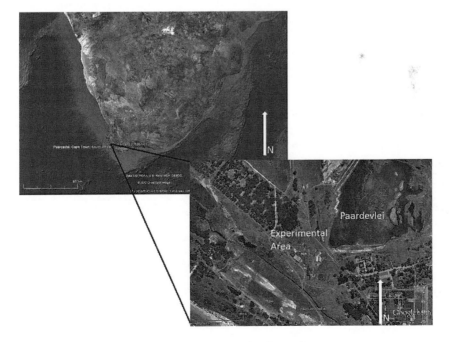

Figure 15.1 Location map for the study site.

to obtain information about the site. The area to be remediated was situated where an ammonium-nitrate warehouse was previously operating. The area was approximately 3.6 ha in size. The surface elevation was approximately 6.5 m amsl. A water body (Paardevlei, Figure 15.1) is situated adjacent to the site and covers an area of approximately 50 ha (SRK, 2006). Natural drainage channels to the lake have been canalised away from it and over the years the walls have been raised. The site was situated on coastal plain underlain by Malmesbury Shale formation. Malmesbury rocks, comprised of dark green-grey shale, hornfels and quartzite and they are generally well fractured. Fractures, particularly in the vadose zone, are generally filled with very stiff, dark green-grey clay. The site occurs in the interface zone between the Coastal Plain and Coastal Dune Belt. Quaternary sediments of the Langebaan geology occur at the site.

Somerset West normally receives about 568 mm of rain per year and because it receives most of its rainfall during winter it has a Mediterranean climate. It receives the lowest rainfall (10 mm) in February and the highest (96 mm) in June. The average midday temperatures for Somerset West range from 16.2°C in July to 26.1°C in February. The region is the coldest during July when the mercury drops to 7.2°C on average during the night. The majority of the lowland area in Somerset West is heavily transformed by agriculture and urbanisation and there is minimal natural vegetation remaining but rather predominantly alien herbaceous vegetation (Holmes, 2002). The AECI in particular have mainly bluegum (*Eucalyptus globulus*) trees as well as grasses across the site. A general soil profile (SRK, 2006) for the study site is presented in Table 15.1.

Sampling of groundwater and soil was done and water-level depth measurements were taken prior to installation of the treatment zone. Sampling and analyses of nitrate were carried out to establish the distribution of nitrate concentrations across the area where the nitrate removal would be tested. Distances between points were measured and water-level data were used to construct contours of groundwater-level data and to determine groundwater-flow vectors.

Groundwater nitrate concentrations as N and winter groundwater levels together with the distances between boreholes and position of carbon source tank are shown in Figure 15.2. Water-table elevation data were then interpolated and used to determine groundwater flow at the site based on groundwater-elevation gradients. Figure 15.3 shows the groundwater interpolated elevation map. Groundwater flow is from a high elevation toward a lower elevation. Figure 15.3 shows that flow occurs towards the coastal dune area just east of BH9, and measures were put in place to divert or channel

Table 15.1 Generalised soil profile for the AECI site, Somerset West, after SRK (2006).

Strata	Depth (m bgl)	Description
Surface	0–0.4	Grassed surface with red gravel and ash fill
Transported/alluvium	0.4–1.5	Slightly moist, brown, loose, fine-medium SAND
Transported/alluvium	1.5–2	Moist, light brown sandy CLAY
Transported/alluvium	2–3	Moist – wet, light brown to grey, clayey SAND with fine-calcrete gravel concretion and medium rounded gravel. Groundwater seepage encountered at this depth.

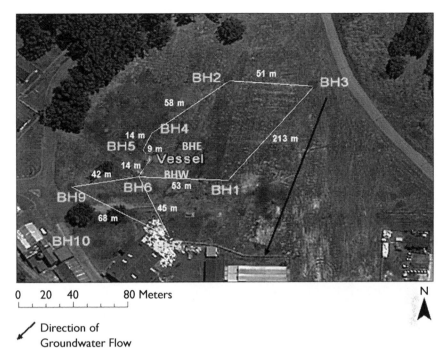

Figure 15.2 Borehole positions and proximity from each other. Groundwater flow direction indicated with black arrow.

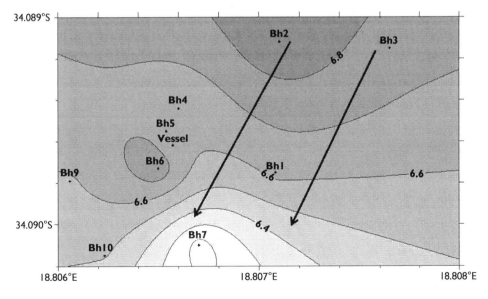

Figure 15.3 Water-level elevation at the experimental site (contours represent groundwater elevation in m). Arrows indicate groundwater flow direction at the site.

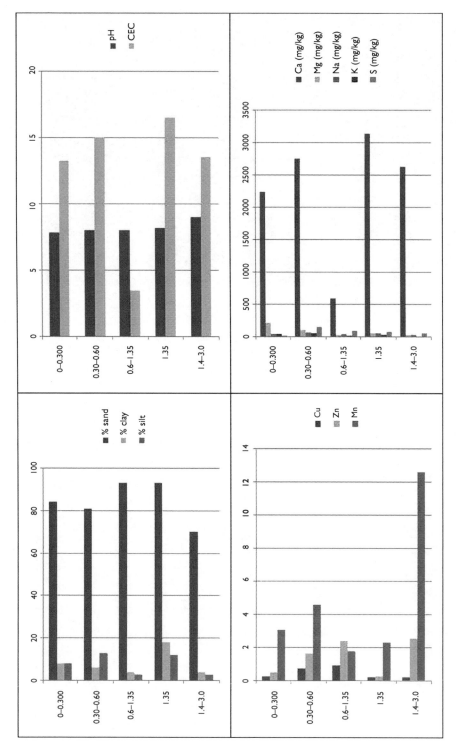

Figure 15.4 Results of soil analyses. These include upper left: particle size analyses; top right: pH and cation exchange capacity; bottom left: metals; and bottom right: exchangeable cations.

flow away from the Paardevlei. Local flow towards the boreholes may be occurring; see arrows indicating groundwater flow direction in Figure 15.3. The carbon source tank was placed between BH5 and BH6. Soil samples were collected during excavation. Figure 15.4 shows the results of soil tests.

Most of the soil profile is composed of predominantly sand-sized particles, with the soil depth of 1.4 m to 3 m having the highest clay content of 10%. This relates to a higher cation exchange capacity for this depth as more charged surfaces are available in clay particles. The predominant soil texture of the aquifer was sandy and porosity is estimated to be about that of sand – i.e. 20% to 35% based on the particle sizes present.

Trace metals copper (Cu), zinc (Zn) and manganese (Mn) were analysed for as it is documented that the presence of metals can enhance denitrification rates at various concentrations (Labbé *et al.*, 2003). The groundwater level was at 1.35 m below surface at the area where excavating took place. It varied across the site. Results show that manganese is the dominant trace metal at most soil depths and it is particularly concentrated within the saturated zone below 1.35 m depth, while zinc is more concentrated just above the water table. Manganese oxides are among the strongest naturally occurring oxidizing agents in the environment, having high sorptive capacities and participating in various redox reactions with both organic and inorganic compounds.

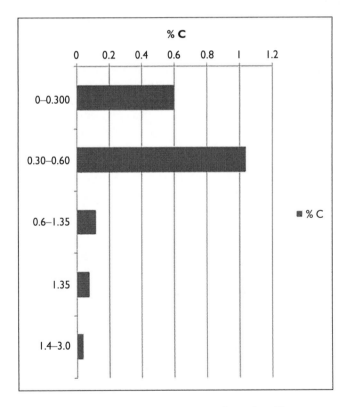

Figure 15.5 Percentage carbon in the soil profile.

Exchangeable nutrients were analysed as an indication of availability in the soil environment. The results of the cations or exchangeable nutrients show that calcium is the dominant cation in the soil at all depths. The presence of calcareous boulders and calcrete nodules at some depths as well as the fact that the position of the excavation was at a contact between more clayey type geology and coastal calcareous sands supported the fact that calcium was the dominant cation at all soil depths. Sulphur is a redox sensitive species which may be affected by oxidation reduction processes or condition of the subsurface. Figure 15.5 shows the percentage carbon along the soil profile. This was tested to evaluate the possible occurrence of natural denitrification.

The percentage Total Organic Carbon (TOC) results show that the maximum total organic carbon is just over 1% in the groundwater. This would result in a low likelihood for natural denitrification at the site, hence it was deemed necessary to amend the subsurface with a slowly degradable carbon source.

15.3 FIELD IMPLEMENTATION AND RESULTS OF MONITORING AT THE SITE

Results from previous laboratory experiments (Israel, 2007; Israel *et al.*, 2009) were used as an indication of treatability and rates of reactions for the selected carbon source. The barrier or wall thickness as well as concentration or mass of carbon source required for a particular nitrate concentration was determined using stoichiometric relationships for denitrification as well as results from a laboratory flow-through experiment. Calculations took into consideration an average or potential flow rate based on soil type, size of the area to be treated as well as desired resultant nitrate concentration. The selected carbon source was chosen based on availability, treatability, costs, effectiveness and minimal nuisance products or side reactions during laboratory experiments as well as longevity of the carbon source which would in turn reduce the need or frequency for maintenance at the site. Untreated woodchips were freely available for the experiment on site. Five thousand kilograms of woodchips were used; this was limited by the size of the treatment vessel (tank) used. The determination of the mass of a litre of sawdust was calculated via a mini experiment. It was determined that the mass to volume ratio for sawdust was roughly 1:1. The calculated requirement for sawdust was between 4 980 kg and 6 350 kg which compared favourably with the amount that could be accommodated in the tank. Photos were taken throughout the implementation phase. Figure 15.6 shows the installation of the treatment vessel.

The tank was fully slotted to allow flow through the treatment material and into the aquifer. The upper left picture in Figure 15.6 shows the start of the digging process, while the lower left picture represents the tank used during the experiment. The upper right picture shows the tank in place within the groundwater table, while the lower right picture shows the fully slotted PVC pipes used as monitoring points to track trends of groundwater quality within the tank and adjacent to the tank. The tank used was a 5 000 l tank with a height of 1.37 m, and a diameter of 2.15 m. Monitoring of the central point in the tank, the two adjacent points as well as already existing boreholes continued for 2 d after the tank was in place. Thereafter, biweekly

Figure 15.6 Digging and emplacement of tank used for carbon source during the experiment. Top left shows the start of the excavation; bottom left shows the preparation of the tank. Top right shows the emplacement and filling of the tank. Bottom right shows placement of PVC for monitoring points additional to existing boreholes at the site.

site visits took place. Figure 15.7 shows some results of nitrate concentrations for the tank and the two points adjacent to the tank.

The boreholes adjacent to the tank had initial nitrate concentrations of 320 mg/l (labelled BHE on Figure 15.2) and 280 mg/l (labelled BHW on figure 15.2) on the first day of sampling (Figure 15.7). On the second sampling trip (2 weeks after), both boreholes showed a considerable decrease in their nitrate concentration, with the East showing a 51.6% decrease in concentration to 155 mg/l and the West a 96% removal of nitrate to 11 mg/l. Within 2 months, which equates to 4 sampling events, all the nitrate was reduced in the boreholes. Figure 15.7 shows that the nitrate concentration was below 1 mg/l in the tank from day 2 of installation. The nitrate levels remained low for the remainder of the experiment in all these boreholes until the next rainfall event.

Figure 15.8 shows the surface interpolation of nitrate concentrations on the site before the tank was in position as well as the nitrate and sulphate concentrations one month after the tank was in place. Figure 15.9 shows the borehole logs for BH5 and BH6, the closest drilled boreholes to the tank. Table 15.2 shows the depth and seasonal water levels for boreholes at the site.

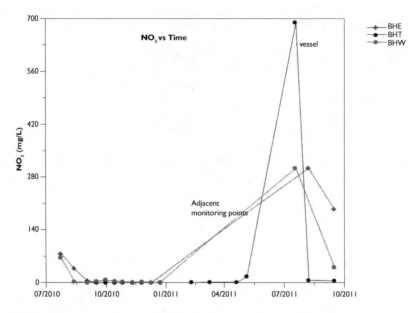

Figure 15.7 Nitrate concentration mg/l as NO₃ in the tank and two adjacent boreholes from July 2010 up to October 2011.

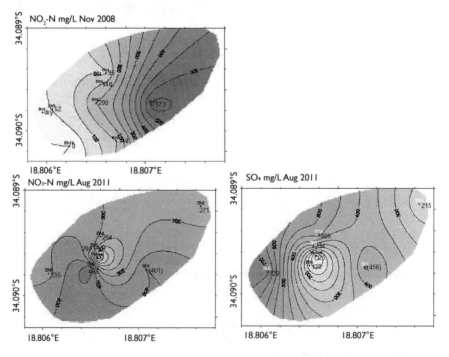

Figure 15.8 Surface interpolation of site nitrate concentrations before installing the tank/vessel as well as plots showing concentrations of nitrate and sulphate one month after installation.

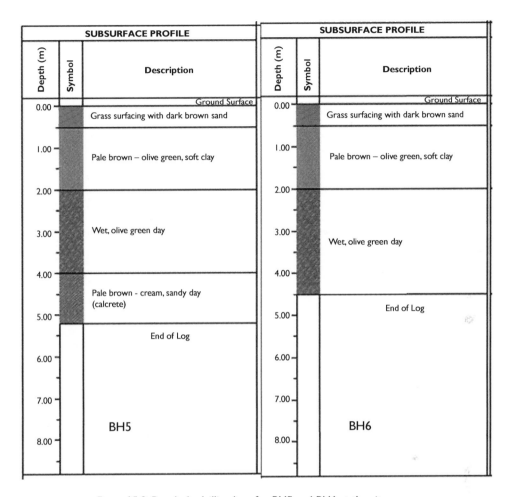

Figure 15.9 Borehole drilling logs for BH5 and BH6 at the site.

Table 15.2 Available details of boreholes on the site.

Borehole	BH4	BH5	BH6	BH7
Drilled depth	6 m	5.5 m	4.5 m	4.5 m
Rest WL	1.5 m	1.5 m	1.5 m	1.5 m
Winter WL	1.5 m bgl	2.36 m bgl	2.6 m bgl	2.4 m bgl
Summer WL	3.24 m bgl	3.47 m bgl	3.17 m bgl	No data available
BH diameter	63 mm	63 mm	63 mm	63 mm
Clay depth	0.5 m–4 m	0.5 m–4 m	0.5 m–4.5 m	0.5 m–4.5 m
Sandy clay depth	4 m–6 m	4 m–5.5 m	Not penetrated	Not penetrated

Data in Table 15.2 were not available for the site prior to emplacement of the vessel as monitoring of the site occurred once per year. Summary plots of parameters per borehole (NO_3^- SO_4^{2-}, and EC) were done for boreholes BH4 (Figure 15.10), BH5 (Figure 15.11) and the two boreholes adjacent to the tank BHE (Figure 15.12) and BHW (Figure 15.13).

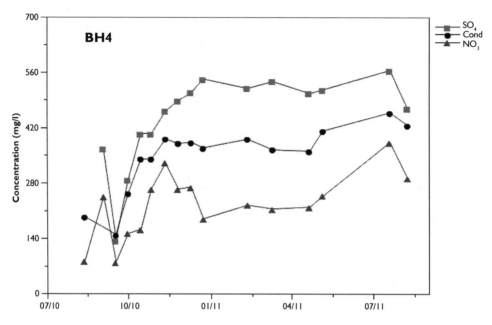

Figure 15.10 Time series plot of NO_3, SO_4 and EC at BH4 from July 2010 up to October 2011.

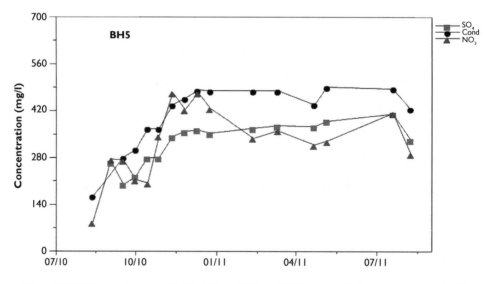

Figure 15.11 Time series plot of NO_3, SO_4, and EC at BH5 from July 2010 up to October 2011.

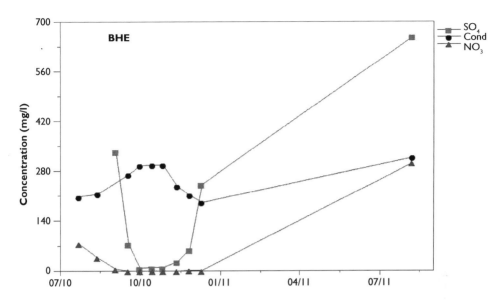

Figure 15.12 Time series plot of NO$_3$, SO$_4$ and EC at BHE from July 2010 up to October 2011.

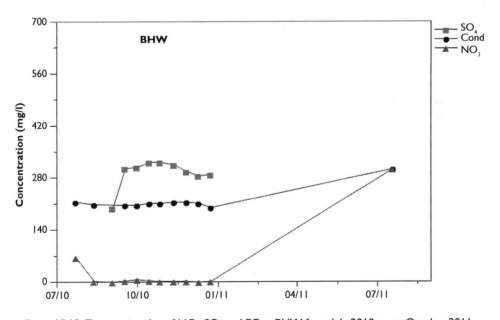

Figure 15.13 Time series plot of NO$_3$, SO$_4$ and EC at BHW from July 2010 up to October 2011.

15.4 DISCUSSION

The aim of the experiment was to evaluate whether denitrification of high industrial-level nitrate concentrations would occur, using sawdust, and to determine how effective the sawdust would be in denitrifying. The lifespan of sawdust as a carbon source was also under investigation. The mass in kilograms of sawdust required to effect one year's denitrification was calculated for this experiment using stoichiometric ratios. Denitrification takes place using immediately available carbon sources, hence breakdown products of sawdust (e.g., tannins and cellulose/acetic acids) which are likely to be easier to use by bacteria will take part in the reaction (Robertson and Cherry, 1995; Robertson *et al.*, 2003). Results thus far showed that the carbon requirement calculated and sawdust load used was effective beyond the period of 365 d, which could be due to the slow release of available carbon from sawdust; this is in agreement with work done by Robertson and Cherry, 1995; Robertson *et al.*, 2003; as well as Schipper and Vojdovic-Vukovic, 2000; 2001; and Shipper *et al.*, 2004.

Since the study area contained calcrete boulders in places as well as clay lenses, careful attention was given to the tank holding the sawdust. Complete slotting was used to ensure that flow would occur toward the tank from the surrounding saturated zone. Hydraulic conductivity has been pointed out as an important factor in many studies (Schipper *et al.*, 2004; Schipper and Vojvodic-Vukovic, 2000; 2001; Robertson and Cherry 1995; Robertson *et al.*, 2000; Barkle *et al.*, 2008). Excavation and emplacement of the barrier was done bearing in mind that the hydraulic conductivity of the tank and sawdust needed to be greater than that of the water-bearing lithologies to ensure flow towards and through the tank. The water-bearing lithologies contained a mixture of course sand and clay material, hence porosity ranged between 25% and 60%. Even though the porosity of clays can be high, the permeability is not. The course sand present in the subsurface would most likely be preferred by fluid movement; the hydraulic conductivity of sand ranges between 10^{-2} and 10^{-7} m/s.

The removal of initial nitrate levels in the vicinity of the tank takes place within the first 2 days; this finding compares favourably with the work of Turner and Patrick (1968) which shows that nitrate is already removed within 2.5 d. Boreholes further away from the tank showed a decline in nitrate concentration approximately 2 months and 4 months after the tank was put in place, and then showed an increase. The boreholes are approximately 10 m and 30 m away from the tank. The percentage nitrate removal was 40% in BH5 (Figure 15.11) and 45% for BH4 (Figure 15.10) prior to a subsequent increase in concentration. It is envisaged that a diluted and partially consumed dissolved organic carbon concentration compared to the concentration in the tank may have reached these points. It appears that preferential flow took place even though the two closest monitoring boreholes were more or less equidistant from the tank. The borehole logs showed that the boreholes in the preferred direction of greater flow velocity are drilled through the main water-bearing lithology on the site composed of sand and clay, while other boreholes are drilled into clay which has slower flow velocities.

Higher initial manganese levels in the BHE borehole may have contributed to the rapid reduction of both nitrate and sulphate, together with the highly reduced conditions produced by the strong oxidising agent. This finding is in agreement with results reported in an experiment conducted by Labbé *et al.* (2003), who investigated the effects of the addition of trace metals on denitrification rate. The delayed response in boreholes further from the tank is due to the presence of low permeability particles within the aquifer, which delayed the flow of elevated dissolved organic carbon from reaching boreholes further away. The slow natural flow rate at the site was the rate-determining part of the reaction in the subsurface.

15.5   LIMITATIONS OF THE SYSTEM

The tank used was only 1.37 m in height while the winter water table was at a depth of about 1.4 m below ground level (bgl) and the summer water table depth ranged from 1.8 m (minimum depth) to over 3 m (maximum water-table depth) in some boreholes. This means that the tank's location was limited to between 2.8 m bgl and 1.48 m bgl. Borehole logs for BH5 and BH6 show the lithologies at the site (Figure 15.9).

Uncertainty with respect to the ability to sample all points during summer, as well as summer reaction rates or occurrence exist. Some boreholes have water levels of below 2.8 m during summer. The bedrock was not encountered during the excavation hence there is uncertainty about underflow in the treatment zone as well as flow by-passing the treatment zone. The presence of disconnected clay lenses as well as calcareous sands in certain parts of the aquifer presented potential for areas or zones of variable flow. Monitoring was done bi-weekly using indicator parameters to accommodate the available budget; this limited the amount of data obtainable for the study area. Uncertainty with respect to distribution of clay lenses and calcrete boulders presented some limitation with respect to the local flow direction and velocity at the site. The boreholes adjacent to the tank had initial nitrate concentrations of 320 mg/l (labelled BHE on Figure 15.2, also see Figure 15.12) and 280 mg/l (labelled BHW on Figure 15.2, also see Figure 15.13) on the first day of sampling (Figure 15.7). On the second sampling trip (2 weeks after), both boreholes showed a considerable decrease in their nitrate concentration, with the BHE (Figure 15.12) showing a 51.6% decrease in concentration to 155 mg/l and the BHW (Figure 15.13) a 96% removal of nitrate to 11 mg/l. Within 2 months, which equates to 4 sampling events, the nitrate levels in all the boreholes had been reduced. The presence of clay is not ideal, but not limiting either. It simply means that flow is much slower than desired in some areas of the site, and hence the denitrification will take place over a longer period in those areas.

15.6   CONCLUSIONS

It was concluded that *in situ* field denitrification successfully removed nitrate within a short period of time, in this case about one month for an area of approximately up to 2.5 m away from the centre of the tank. Proximity to the main denitrification zone, as well as ease of flow path toward other boreholes and effects of infiltration, dilution, dissolution or replacement on soil particles as well as competing reactions of nitrification and denitrification all affect the results seen at a specific point in time. It was concluded that the tank's carbon source reached boreholes BH4, BH5 and BH6 during the course of the experiment; however, it may have been in insufficient concentration to effect total denitrification at these points. The presence of trace metals on soil surfaces and within the saturated zone may have enhanced the denitrification and nitrification at some boreholes. Sulphate reduction occurred due to nitrate already being used up in BHE (Figure 15.12). This is explained by the redox sequence in waterlogged soils and the saturated zone. When sulphate is available in the sub-surface while carbon is still available as a reducing agent, and reducing conditions prevail, sulphate reduction will occur. An important issue that came across from reviewing the methods used is that budgets play a big role in the construction and implementation time and sophistication. It was successfully demonstrated that *in situ* denitrification can effect total removal of nitrate of high industrial-level nitrate concentrations using sawdust as a cheaply available and slowly degradable carbon source. Despite the presence of clay, the reaction proceeded successfully. Continual monitoring will enable actual rate calculation for points further away from the tank. The technology can be implemented cheaply, and proved feasible for removing nitrate from the area around the tank within 2 months. Continual monitoring of the site is ongoing.

15.7 RECOMMENDATIONS

To test such or similar technologies in the field, site characterisation is essential as every site has its own unique prevailing conditions. This includes a minimum of testing soil and groundwater physical and chemical properties, flow direction and flow rate, if possible. Ideally, the site should be monitored prior to implementation to determine flow directions and rates; especially taking changes in flow direction and rate into account during seasonal changes. It is also important to test for nitrate and/or other indicator parameters for the specific site to understand the distribution of pollutants at the site prior to selecting the most suitable area to place the treatment barrier. Historical data for any site can give insight regarding the pollution or spill type events, water-level trends, chemistry, pollution or clean-up initiatives that have been done at the site. Treatability studies using material from the site may prove invaluable during field implementation.

ACKNOWLEDGEMENTS

The project team would like to acknowledge the Water Research Commission and the CSIR for funding the research as well as AECI-Heartland Leasing who agreed to the use of their site for testing of the method and their assistance in field implementation.

REFERENCES

Addiscott, T.M. & Benjamin, N. (2004) Nitrate and human health. *Soil Use and Management,* 20, 98–104.
Barkle, G.F., Schipper, L.A., Burgess, C.P. & Painter, B.D.M. (2008) *In situ* mixing of organic matter decreases hydraulic conductivity of denitrification walls in sand aquifers. *Ground Water, Monitoring and Remediation* 28, 57–64.
Canter, L.W. (1997) *Nitrates in Groundwater.* New York, NY, Lewis Publishers/CRC Press.
Department of Water Affairs and Forestry (DWAF) (1996) *South African Water Quality Guidelines, 2nd edition, Volume 5: Agricultural Use: Livestock Watering.* Pretoria, South Africa, Department of Water Affairs and Forestry.

Department of Water Affairs and Forestry (DWAF) (1998) *Guidelines for Domestic Water Supplies*. Pretoria, South Africa, Department of Water Affairs and Forestry.

Gavascar, A.R. (1999) Design and construction techniques for permeable reactive barriers. *Journal of Hazardous Materials* 68, 41–71.

Holmes, P. (2002) Appendix 2: Vegetation, specialist study on the potential impact of the proposed N1/N2 winelands toll highway project on the affected vegetation and plant species. Submitted to Crowther Campbell & Associates on behalf of South African National Roads Agency Limited, South Africa.

Innovative Technology (2000) Summary report: *in situ* redox manipulation, subsurface contaminants focus area. Prepared for US Department of Energy, Office of Environmental Management and Office of Science and Technology, Washington, DC, USA.

Israel, S. (2007) *In situ* denitrification of nitrate rich groundwater in Marydale, Northern Cape. MSc thesis, Faculty of Agriculture, Department of Soil Science, Stellenbosch University, Stellenbosch, South Africa.

Israel, S., Tredoux, G., Fey, M.V. & Campbell, R. (2005) Nitrate removal from groundwater and soil. In: *Proceedings of the Biennial Ground Water Conference, 7–9 March 2005, CSIR International Convention Centre, Pretoria, South Africa*.

Israel, S., Engelbrecht, P., Tredoux, G. & Fey, M.V. (2009) *In situ* batch denitrification of nitrate rich groundwater using sawdust as a carbon source – Marydale, South Africa. *Water Air and Soil Pollution*, 204, 177–194. [Online] Retrieved from: doi: 10.1007/s11270-009-0036-6.

ITRCWG (2000) Technology overview, emerging technologies for enhanced in situ bioremediation (EISBD) of nitrate-contaminated ground water. Interstate Technology and Regulatory Cooperation Work Group, enhanced *in situ* bioremediation work team, Washington DC, USA.

Labbé, N., Parent S. & Villemur, R. (2003) Addition of trace metals increases denitrification rate in closed marine systems. *Water Research* 37, 914–920.

Mateju, V., Cizinska, S., Krejci, J. & Jonach, T. (1992) Biological water denitrification: A review. *Enzyme Microbial Technology* 14, 170–183.

Mercado, A., Libhaber, M. & Soares, M.I.M. (1988) *In situ* biological groundwater denitrification: Concepts and preliminary field test. *Water Science Technology* 20, 197–209.

Robertson, W.D. & Cherry, J.A. (1995) *In situ* denitrification of septic system nitrate using reactive porous media barriers: Field trials. *Groundwater* 33, 99–111.

Robertson, W.D., Blowes, D.W., Ptacek, C.J. & Cherry, J.A. (2000) Long-term performance of *in situ* barriers for nitrate remediation. *Groundwater* 38, 689–695.

Schipper, L.A. & Vojvodic-Vukovic, M. (2000) Nitrate removal from groundwater and denitrification rates in a porous treatment wall amended with sawdust. *Ecological Engineering*, 14, 269–278.

Schipper, L.A. & Vojvodic-Vukovic, M. (2001) Five years of nitrate removal, denitrification and carbon dynamics in a denitrification wall. *Water Research* 35, 3473–3477.

Schipper, L.A., Barkle, G.F., Hadfield, J.C., Vojvodic-Vukovic, M. & Burgess, C.P. (2004) Hydraulic constraints on the performance of a groundwater denitrification wall for nitrate removal from shallow groundwater. *Journal of Contaminant Hydrology* 69, 263–279.

SRK Consulting (2006) *AECI, Somerset West, Ammonium nitrate warehouse characterisation report*. Somerset West, South Africa, Heartland Leasing (Pty) Ltd. Report No 368387/1.

Tredoux, G., Engelbrecht, J.F.P. & Talma, A.S. (2001) Nitrate in groundwater in Southern Africa. In Seiler & Wohnlich (eds.) *New approaches characterizing groundwater flow*. Lisse, Netherlands, Swets & Zeitlinger Publishers. pp. 663–666. ISBN 90651 848 X.

Turner, F.T. & Patrick, W.H. (1968) Chemical changes in waterlogged soils as a result of oxygen depletion. *Transactions of the 9th International Congress of Soil Science (Adelaide, Australia)*, 4, 53–65.

Chapter 16

Classification of surface-water – groundwater interaction of the Mokolo and Lephalala River systems in the Limpopo WMA

T.G. Rossouw[1], M. Holland[2] & N. Motebe[3]
[1]SLR Consulting, Menlyn, Pretoria, Gauteng, South Africa
[2]DeltaH Consulting, Pretoria, Gauteng, South Africa
[3]Department of Water Affairs, Pretoria, Gauteng, South Africa

ABSTRACT

To be able to quantify the groundwater component of the Reserve, the groundwater volumes needed for basic human needs and ecological water requirements need to be quantified. The concept of the Groundwater Reserve and its focus on sustaining rivers and wetlands is not well developed. As a result, water-resource managers in the past decade have put the spotlight on hydrological issues such as groundwater – surface-water interaction. The aim of this investigation was to classify parts of the Mokolo River and lower Lephalala River. A simple classification scheme based on the geology and the hydraulic gradients between surface water and groundwater was used to characterise potential groundwater-surface water interactions. The study showed a clear interaction with the alluvial aquifer whereas a more complex interaction between the river bed and the host rock was identified. In this case a number of situations can occur, namely if the flux is high enough, groundwater will enter the river as baseflow, and if the flux is not high enough it could be used by riparian vegetation or enter the river at selected places. Classification of the river system can be regarded as one of the most important step in quantifying surface-water – groundwater interaction.

Keywords: Mokolo River system, Lephalala River, hydraulic gradient, surface-water – groundwater interaction, groundwater chemistry

16.1   INTRODUCTION

In ephemeral river systems, groundwater flow in the river bed (i.e. within the alluvial sediments) along with regional groundwater flow towards the river bed is often the main water sources which sustain isolated, but ecologically important pools and river stretches. Hence it is of utmost importance to consider groundwater flow both within the alluvium and within the underlying bedrock (country rock) adjacent to the river in characterising the groundwater contribution to the Ecological Reserve. A simple classification scheme based on the geological setting of selected river stretches and the prevailing hydraulic gradients between surface water and groundwater, as proposed by Dennis (2010), was used to characterise potential groundwater – surface-water interactions. The aim of this investigation was to classify parts of the Mokolo River in the vicinity of the 'Shot Belt' pool and a 5 km stretch, the 'Villa Nora' river stretch, of the lower Lephalala River system in Limpopo Province, South Africa, to assist in the evaluation of groundwater – surface-water interaction (Figure 16.1).

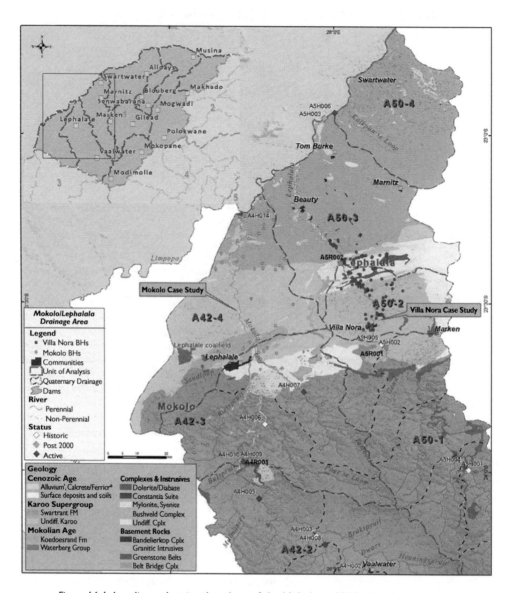

Figure 16.1 Locality and regional geology of the Mokolo and 'Villa Nora' case studies.

16.2 METHODOLOGY

A simple classification scheme (Figure 16.2a and Figure 16.2b) based on the geological setting of selected river stretches and the prevailing hydraulic gradients between surface water and groundwater, as proposed by Dennis (2010), will be used to characterise potential groundwater – surface-water interactions. The aim

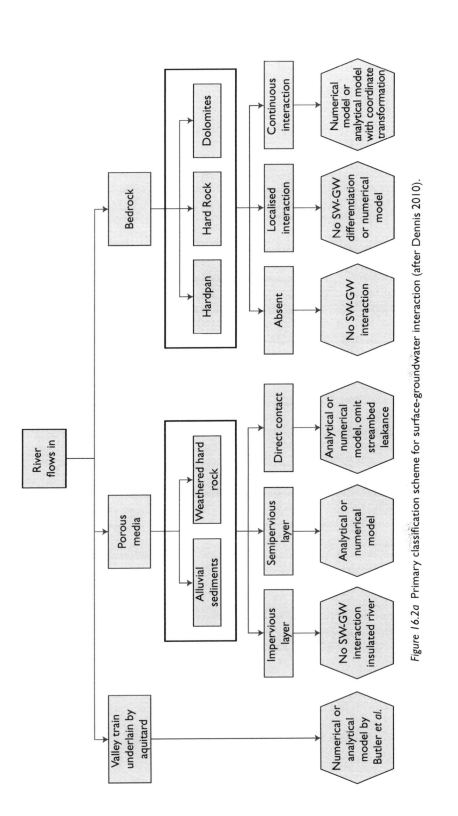

Figure 16.2a Primary classification scheme for surface-groundwater interaction (after Dennis 2010).

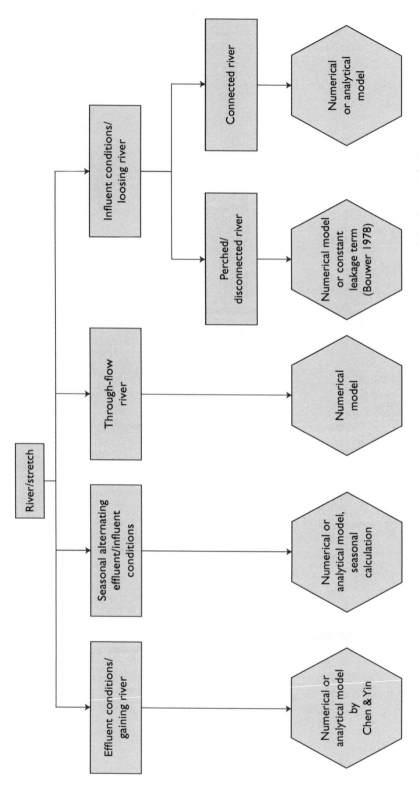

Figure 16.2b Secondary classification scheme for surface-groundwater interaction (after Dennis 2010).

of this investigation was to classify parts of the Mokolo River in the vicinity of the 'Shot Belt' pool and a 5 km stretch, i.e. 'Villa Nora' river stretch, of the lower Lephalala River system to assist in the evaluation of groundwater – surface-water interaction. Geophysical investigations employing electromagnetic (EM) and magnetic (MAG) techniques to identify anomalous zones associated with deeper weathering and fracturing that could act as conduits for groundwater storage and movement together with borehole hydrographs (water-level monitoring) help in die classification and behaviour of the alluvial aquifer systems.

16.3   CASE STUDY

Two areas of investigation (i.e. 'Shot Belt' pool in the Mokolo River and 'Villa Nora' river stretch in the lower-Lephalala River) were identified for a surface-water – groundwater interaction study. The 'Shot Belt' pool site is located approximately 10 km north of Lephalale towards Groblers Bridge (i.e., a border post) next to the Mokolo River on the farm Shot Belt and comprises catchment A42 (23°33'0.9"S; 27°42'47"E). The 'Villa Nora' river stretch site locality is located approximately 30 km west of Marken and comprises catchment A50. The study areas fall within the Limpopo Water Management Area (WMA) forming part of the Limpopo Province in the northern part of the Republic of South Africa. The geomorphology varies depending on the type of geological formation present. The southern parts, origin of the Mokolo and Lephalala Rivers, are typically characterised as relatively mountainous areas (Waterberg Mountain Range) where the geomorphology changes into more flat-lying areas towards the north (underlain by Karoo Supergroup, Bushveld Igneous Complex and basement rocks) at the confluence of the Limpopo River with elevations ranging from 600 m amsl to 800 m amsl.

16.3.1   Geology of the study areas

16.3.1.1   Basement complex (Limpopo Mobile Belt)

The northern extent of the study areas is underlain by rocks of the Beitbridge Complex and Hout River Gneiss formation forming part of the Basement Complex. According to Barnard (2000) and Foster (1984), the Basement Complex is mainly characterised by granites, granodiorite, migmitites and gneisses but also includes metamorphosed sediments, slate, talc schist and sandstone. These Archaean crystalline basement rocks are the oldest rock formations in the study area and are dated at about 3.1 Ga.

16.3.1.2   Bushveld Igneous Complex (BIC)

Intrusive igneous rocks are found within the central-Lephalale River stretch of the study area belonging mainly of the Lebowa granite suite. These 2.06 Ga old rocks range in composition from ultramafic to acidic and consist mainly of granodiorite, gabbro, norite, anorthosite and granite. Typical intergranular and fractured-rock aquifer systems can be expected.

16.3.1.3 Waterberg formation

The southern part of the study area is underlain by the Waterberg Formation that consists of three main subgroups, namely the Setlaole Formation, Makgabeng Formation and Mogalakwena Formation. The basal Setlaole Formation rests non-conformably on the basement rocks, and is composed of coarse granulestone and is locally conglomeratic. The Makgabeng Formation is deposited conformably on the Setlaole Formation, and consists of large-scale trough and planar cross-bedded fine- to medium-grained sandstone, deposited as an aeolianite (Bumby, 2000). The Mogalakwena Formation, in contrast, lies disconformably above the Makgabeng Formation. These strata consist of interbedded sheets of granulestone and conglomerate in proximal areas, grading into trough cross-bedded granulestones and sandstones in distal areas to the southwest (Bumby, 2000).

16.3.1.4 Karoo supergroup

The Karoo Supergroup, formed during the Palaeozoic and Mesozoic Eras (545 Ma–65 Ma), underlies the central-Mokolo River Stretch and consists mainly of sedimentary rocks. The Waterkloof Formation (Ecca Group), forming part of the Ellisras basin, comprises diamictite, mudstone and conglomerates and rests unconformably on the Waterberg Formation (Brandl, 1996).

16.3.1.5 Quaternary and alluvial deposits

Quaternary deposits cover large portions of the Basement Complex (Limpopo Mobile Belt) and the northern reaches of the Waterberg Formation. Sediments such as calcrete, ferricrete, gravel, red sand and alluvium are expected. Alluvium of up to 5 m thick with a coarse sand base is present along the river stretches and forms an important local aquifer especially during the summer (rainfall) seasons.

16.3.2 'Shot Belt' pool in the Mokolo River case study

The sub-division of the Mokolo River alluvial aquifer consists of two different sections based on the aquifer properties determined by Vipond (1988), WSM (1999) and Groundwater Resource Direct Measure (GRDM) resource-classification system. The following properties were attributed to four sections of the alluvium associated with the Mokolo River:

- Mokolo A42-3 is characterised by a mountainous region underlain by sediments of the Waterberg Formation with deep eroded valleys and cliffs. Bases on the National Groundwater Achieve (NGA) the average depth of the alluvial sands in the Mokolo River is 12.4 m with an average transmissivity (T) value of 1 489 m²/d. The average hydraulic conductivity (k) of the alluvial sands is 175 m/d. The hydraulic gradient (i) towards the river bed from the surrounding Waterberg Formation is 0.00073, indicating a steeper gradient compared to the other sections.
- Mokolo A42-4 is characterised by alluvium underlain by rocks of the Karoo Supergroup (Ecca Group). The alluvium has an average depth of 7.5 m. The

transmissivity (T) is approximately 999 m²/d with an average hydraulic conductivity (k) of 132 m/d. The section is also characterised by alluvium underlain by igneous and metamorphic rocks of the Basement Complex that has an average alluvium depth of 6.5 m. The average transmissivity (T) is 810 m²/d and the average hydraulic conductivity (k) is 117 m/d. The hydraulic gradient (i) towards the river is estimated at 0.00043, again indicative of a more gentle topography.

16.3.2.1 Surface-water – groundwater classification

Based on the surface-water – groundwater classification proposed by Dennis (2010) the two primary geological classification schemes characterise the Mokolo River flowing, i.e. (1) through hard bedrock media characterised by the continuous interaction between surface water and groundwater – Waterberg Formation; and (2) through porous media characterised by semi-pervious alluvial sediments representing the Karoo Supergroup and Basement Complex. The secondary hydraulic classification scheme for the river stretch points to seasonal alternating effluent/influent conditions between the Mokolo River and the alluvial aquifer.

Three geophysical transverse lines were surveyed (Figure 16.3), i.e. one parallel and two perpendicular to the 'Shot Belt' pool. From the EM and MAG techniques no defined lineaments could be identified; however, a relatively good idea of the weathered thickness and possible clay anomalies could be seen. Weathering occurs at levels deeper than 60 m around the river stretch and 'Shot Belt' pool with shallow higher resistivity lenses (clay) with thicknesses of less than 10 m below the surface.

Seven boreholes have been drilled in the vicinity of 'Shot Belt' pool, i.e. two boreholes drilled into the rocks of the Karoo Supergroup (Ecca Group) and the remaining five into the alluvial aquifer system next to the 'Shot Belt' pool. Typical borehole lithological logs (Figure 16.4) in the alluvial aquifer system indicate unconsolidated oxidised sandy soil of between 3 m and 10 m underlain by semi-consolidated clay rich muddy sandstone alternating between 4 and 19 metres and weathered sedimentary rocks.

16.3.2.2 Geochemistry

Groundwater samples (Figure 16.5a and Figure 16.5b) located in the lower-Mokolo River stretch (quaternary catchment A42J) tends to have a more Cl- rich signature with higher TDS values. Groundwater samples of the higher-lying areas, i.e. Waterberg Mountain Range (quaternary catchment A42G) have a HCO_3- enriched groundwater signature with lower TDS values. This is an indication that the Waterberg Mountain Range represents a recharge zone whereas groundwater within the lower-Mokolo River stretch represents a hydro-geochemically-evolved groundwater with varied recharge mechanisms. The Na-Cl facies represent the dominant groundwater type.

Groundwater samples were collected from the 'Shot Belt' pool, inflows to the pool and boreholes in the alluvial and in the Ecca Group. Boreholes located in the alluvium showed the same geochemical signatures compared to the 'Shot Belt'

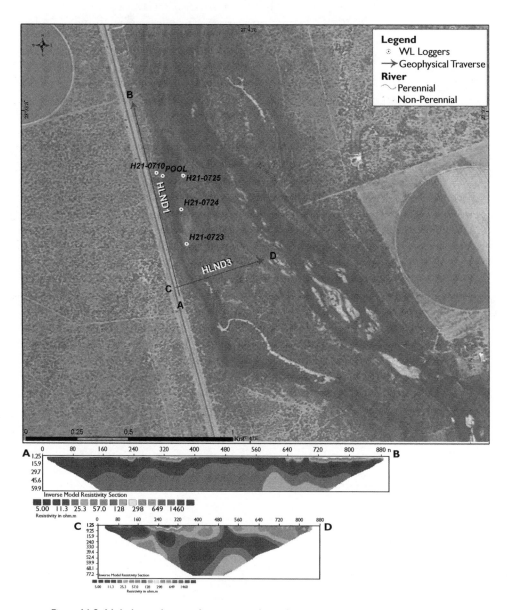

Figure 16.3 Mokolo study area showing geophysical traverses and monitoring boreholes.

pool and inflow stream. Samples drawn from the borehole drilled into rocks of the Ecca Group next to the 'Shot Belt' pool show higher TDS values compared to the other alluvial boreholes as well as a Cl-dominated/enriched anion signature.

The groundwater chemistry of the alluvial aquifer resembles the chemistry of water in 'Shot Belt' pool and inflows to the pool. Based on the single run samples this

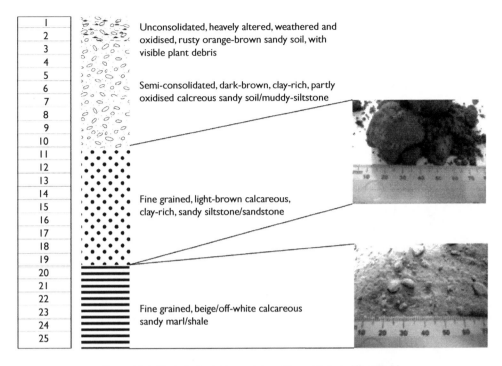

1	Unconsolidated, heavily altered, weathered and
2	oxidised, rusty orange-brown sandy soil, with
3	visible plant debris
4	
5	
6	Semi-consolidated, dark-brown, clay-rich, partly
7	oxidised calcreous sandy soil/muddy-siltstone
8	
9	
10	
11	
12	
13	
14	Fine grained, light-brown calcareous,
15	clay-rich, sandy siltstone/sandstone
16	
17	
18	
19	
20	
21	
22	
23	Fine grained, beige/off-white calcareous
24	sandy marl/shale
25	

Figure 16.4 Typical alluvial log retrieved from Mokolo 'Shot Belt'.

resemblance points to the likelihood of some interaction between groundwater and surface water although the rate of such interaction is probably negligible.

16.3.2.3   Groundwater-level monitoring

As part of an earlier study (Vipond, 1988), between October and November 1987 a total volume of 7.4 Mm³ was released from the Mokolo Dam and the water-level response was monitored in piezometers installed in the alluvium along the entire channel length. This was done in response to the effects of seasonal droughts in the Lephalala district and to increase the viability of irrigation dependent on abstraction from the Mokolo channel and the alluvium. From the water-level recessions the following observations were made: It took approximately 15 d for the volume released from the dam to reach the Limpopo River (with respect to water-level responses in the piezometers) and it took approximately 4 d for the volume released from the dam to reach the 'Shot Belt' pool (with respect to water-level responses in the piezometers).

As part of the study reported here, groundwater-level (Figure 16.6) fluctuations were measured continuously from 21 October 2009 to 21 February 2011 in five of the boreholes (i.e., one in the Ecca Group and four in the alluvial aquifer) as well as of the 'Shot Belt' pool. From the analyses, the groundwater flow is generally away from the 'Shot Belt' pool westwards and eastwards towards the alluvial aquifer and

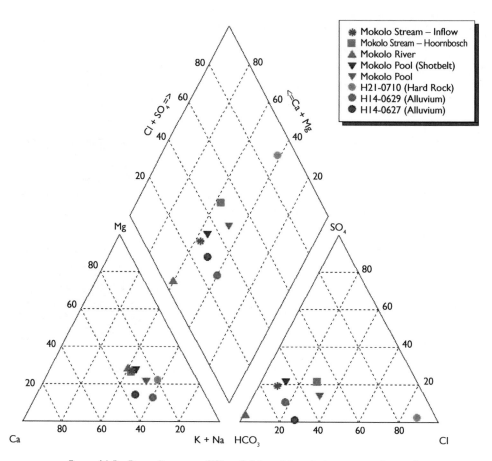

Figure 16.5a Piper diagrams of 'Shot Belt' pool, boreholes next to the pool.

Mokolo River. The groundwater flow indicates influent conditions with respect to the 'Shot Belt' pool and Mokolo River. Generally, rainfall and corresponding changes in weir-flow data (i.e., probably related to spills or scheduled releases from the Mokolo Dam) correlated very well. An overall decrease in weir flows, over the specified period, correlates well with a similar decrease in rainfall data obtained from the Department of Water Affairs (DWA).

Historical reports indicate that water released from the Mokolo Dam is of utmost importance to irrigation projects and probably also to the ecological functioning of the Mokolo River. All water users, especially for irrigation, rely heavily on the releases from the Mokolo Dam indicating that most water from the Mokolo River is driven by surface water and to a small extent from groundwater recharge. Hence the surface-water – groundwater interaction is driven by the release of surface water from the Mokolo Dam as observed in the groundwater fluctuations in the monitored boreholes next to the 'Shot Belt' pool.

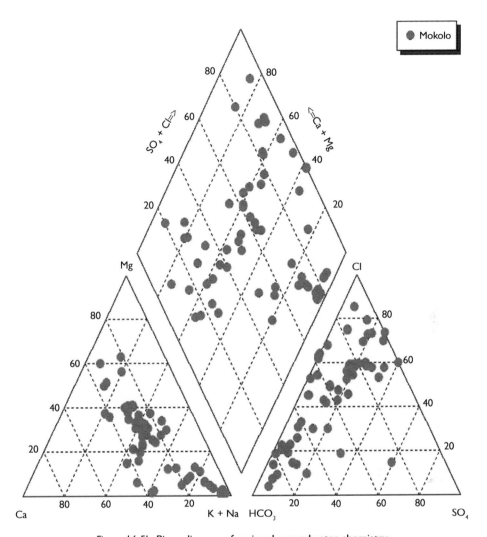

Figure 16.5b Piper diagram of regional groundwater chemistry.

16.3.3 'Villa Nora' river stretch in the lower-Lephalala River case study

Characterising the alluvial aquifer system in the Lephalala River:

Based on the GRDM resource classification the Lephalala River system is classified into four sub-units, namely A50-1, A50-2, A50-3 and A50-4 (Figure 16.1). The area of interest is located within A50-2 and A50-3 (Figure 16.7):

- Lephalala A50-2 is characterised by a discontinuous alluvium layer underlain by rocks of the Bushveld Complex. Where the alluvium occurs it has an average depth of 5 m. The average transmissivity obtained from boreholes drilled within

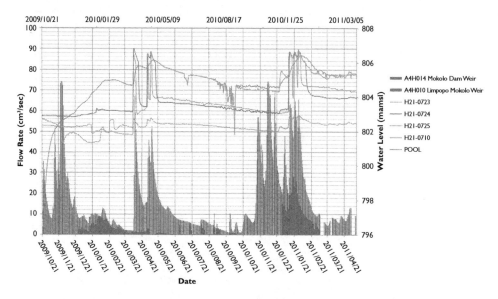

Figure 16.6 Groundwater-level fluctuation presented with Mokolo and Limpopo weir data.

the alluvium within the GRIP framework is 180 m²/d, while average water levels are 3.5 m below ground level (bgl).

- Lephalala A50-3 is characterised by an alluvium layer underlain by rocks of the Beitbridge Basement Complex. The alluvium is slightly thicker compared to A50-2, with an average thickness of about 7 m. Based on the borehole information obtained the transmissivity average is 195 m²/d with average water levels of 5.3 m bgl.

16.3.3.1 Surface-water – groundwater classification

Based on the surface-water – groundwater classification proposed by Dennis (2010) the three primary geological classification schemes characterise the Lephalala River flowing, i.e. (A50-1) through hard bedrock media characterised by the continuous interaction between surface and groundwater represented by the Waterberg Formation; (A50-2) through weathered hard rock in porous media represented by the Bushveld Igneous Complex; and (A50-3) through porous media characterised by semi-pervious alluvial sediments representing the Basement Complex. The secondary hydraulic classification scheme for the river stretch points to seasonal alternating effluent/influent conditions between the Lephalala River and the alluvial aquifer.

Resistivity profiles along and across the Lephalala River at Villa Nora are illustrated in Figure 16.7. The relatively low resistivity (<110 Ωm) is evident in the top layer in the immediate connection with the main river channel. The thickness of the layer varies from a few metres to around 10 m below the present river channel. In places the layer is interrupted by what may be interpreted as an intrusive dyke. Below the stream channel and inferred alluvium layer an indication of high resistivity is visible at depth. This may be interpreted as the Bushveld Complex bedrock. In some

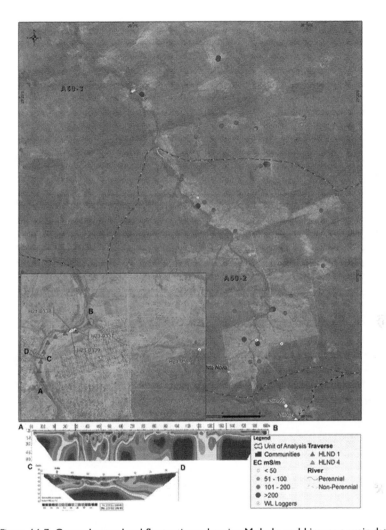

Figure 16.7 Groundwater-level fluctuations showing Mokolo and Limpopo weir data.

sections along the river course the river channel incises into the bedrock with a thin or absent alluvium layer, and in other sections a weathered bedrock aquifer underlies the alluvium. The alluvial aquifers interact with the composite weathered aquifer as well as surface water, both interactions being dependent on, amongst other factors, prevailing groundwater gradients as well as the presence (or lack) of clogging layers in the stream bed resulting in an imperfect hydraulic connection.

16.3.3.2 Geochemistry

Groundwater samples in the 'Villa Nora' river stretch are characterised by the dominance of Mg-Na-HCO$_3$ water facies (Figure 16.8). Some groundwater samples did

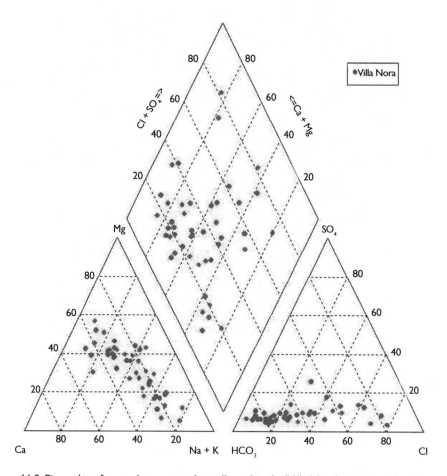

Figure 16.8 Piper plot of groundwater samples collected in the 'Villa Nora' river stretch study area.

show Cl-dominant water facies; however, these are only localised groundwater samples except for a few samples located downstream in the lower Lephalala River. Some groundwater samples did also show a NO_3-dominant anion water facies that can be attributed to anthropogenic processes. The Mg-Na-HCO$_3$ water facies represent the dominant groundwater type in the area.

16.3.3.3 Groundwater-level monitoring

Groundwater-level fluctuations were measured continuously from 5 January 2011 to 13 May 2011 in four boreholes along the Lephalala River. One borehole targeted the Bushveld Igneous Complex, while the other three were located within the Lephalala stream bed and along the river bank at 'Villa Nora' (Figure 16.9). The water table is generally unconfined and local groundwater flow follows the surface drainage towards the alluvial aquifer. Unfortunately no flow-gauging stations were available at the 'Villa Nora' site; however, an upstream gauging station on the Waterberg Formation was used for comparison

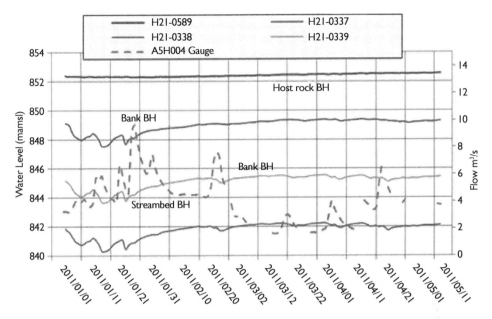

Figure 16.9 Groundwater-level fluctuation and weir data.

purposes. It appears that the water levels in the boreholes are somewhat masked by the storm runoff event observed from the gauging station. Following the storm runoff event a slight dip in the groundwater level is observed before it recovers to pre-storm levels. It is clear that these events have an impact on the water level along the banks of the Lephalala River and more importantly on recharging of the alluvial system. The storm runoff events observed in January had increased the water levels on the river banks by more than 1 m. The borehole within the host rock is much more sensitive to changes in recharge events which in this case may be attributed to the low permeability of the host rock.

16.4   CONCLUSION

The surface-water – groundwater interaction with regard to the water level in the 'Shot Belt' pool is primarily maintained by infrequent overspills from the Mokolo River in the vicinity of the pool in combination with low levels of exchange between the pool and the alluvial aquifer. Shallow clay layers, derived from geophysical investigation (i.e., low resistivity values) and borehole logs, with permeability coefficients of 10^{-3} m/d to 10^{-4} m/d (typical for an aquiclude) were intersected around and underlying the pool. These clay layers significantly limit any potential exchange of water between the pool and the alluvial aquifer. The water-level gradients were consistently away from the pool throughout the monitoring period, indicating losing (i.e., influent) conditions with regard to the pool (i.e., surface water replenishing groundwater). The groundwater chemistry shows a typical regional groundwater flow evolutionary path from higher-lying recharge areas (bicarbonate being the dominant anion) to lower

lower-lying discharge areas (chloride being the dominant anion). To some extent, the groundwater samples indicated interaction between the 'Shot Belt' pool and boreholes drilled in the alluvium.

The characterisation of the surface-water – groundwater interaction at the 'Villa Nora' river stretch indicates that water levels in the boreholes are somewhat masked by the storm runoff event observed from the gauging station. After the runoff event a slight dip in the groundwater level is observed before it recovers to pre-storm levels. It is clear that these events have an impact on the water level along the banks of the Lephalala River and more importantly on recharging of the alluvial system. The geophysical investigation of the 'Villa Nora' river stretch indicates shallow alluvial sediments as the river incised into bedrock or highly weathered bedrock aquifer underlying the alluvial system.

Based on the two case studies, 'Shot Belt' pool and 'Villa Nora' river stretch, surface-water – groundwater interaction is strongly influenced by surface water. Boreholes drilled in the alluvial aquifer in both the Mokolo River and Lephalala River show direct influence from the runoff events either driven by rainfall or surface water released from the dams and only limited influence from groundwater.

REFERENCES

Barnard, H.C. (2000) *An Explanation of the 1:500 000 General Hydrogeological Map: Johannesburg 2526*. Pretoria, South Africa, Department of Water affairs and Forestry.

Brandl, G. (1996) The geology of the Ellisras area. Explanation Sheet 2230. *Geological Survey of South Africa* 1–35.

Bumby, A.J. (2000) The geology of the Blouberg Formation. Waterberg and Soutpansberg Groups in the area of Blouberg Mountain, Northern Province, South Africa, Unpublished PhD thesis, University of Pretoria, South Africa.

Dennis, I. (2010) *Update of the Groundwater Direct Measures Manual – Draft report*. Pretoria, South Africa, Department of Water Affairs.

Foster, M.B.J. (1984) *Preliminary geological report: Steenkoppies and Zwartkrans dolomite compartment*. Director-General of Environmental Affairs, Hydrogeology.

Vipond, S.H. (1988) *The water resource potential of the sand bed of the Mogol River*. Pretoria, South Africa, The Director-General for Water Affairs and Forestry. Technical report No. GH 3570.

Chapter 17

Is the Precambrian basement aquifer in Malawi up to the job?

J. Davies[1,4], *J.L. Farr*[2] & *N.S. Robins*[3]
[1]*British Geological Survey, Wallingford, UK*
[2]*Wellfield Consulting Services, Gaborone, Botswana*
[3]*British Geological Survey, Wallingford, UK*
[4]*University College, London, UK*

ABSTRACT

Repeated failure of borehole supplies in weathered Precambrian basement throughout southern Africa, and in Malawi in particular, has been blamed on a variety of causes. These include poor borehole design, badly sited boreholes, mechanical failure and even implementer ignorance of basic hydrogeological principles. Recent work in Malawi demonstrates that recharge and flow potential of the data-scarce basement aquifer can be investigated by deriving a variety of surrogate and secondary data. Investigation shows that the weathered Precambrian basement aquifer may not always be up to the job of sustaining even modest abstraction sources indefinitely and that many new sources will fail in the short to medium term. Failure is the result of recharge being unable to sustain abstraction within the limited capture zone of an abstraction source, there being little effective lateral inflow of groundwater on the African plateau savannah lands which have very low hydraulic gradients. If lateral groundwater flow does occur across a source capture zone, for example towards the rift-valley escarpment, sources are more sustainable to a degree, depending on the volume of throughflow. Where surface water courses traverse the aquifer, albeit flowing ephemerally or largely as sub-river bed sand rivers, sustainability of borehole sources is increased provided there is hydraulic connectivity to the aquifer. Abstraction from the aquifer within a source capture zone may be as little as 25% of storage in isolated rural areas, more typically about 50% and as high as 75% in the more densely populated areas. If demographic growth requires an increase of just 1% in large areas of Malawi, source failure will occur within a 25-year to 35-year period.

17.1 INTRODUCTION

Weathered and fractured Precambrian crystalline basement provides water for rural populations in 12 of the 15 member states of the SADC community of nations in southern and eastern Africa. Titus *et al.* (2009) describe the typical shallow fractured and weathered basement aquifers that occur within these rocks. Carl Bro, Cowiconsult & Kampsax-Kruger (1980) first described the saprolite profile and its hydraulic properties from work in Tanzania, ideas that were later acknowledged and transferred to Malawi by Smith-Carrington and Chilton (1983).

Reports of borehole failure during repeated major droughts, for example in 1984 and 1992, indicate that the long-term sustainability of the resource may be in doubt. The impact of drought upon groundwater systems is perceived to be less severe and less immediate than upon surface-water systems and groundwater resources form a buffer to mitigate against the impact of such droughts (Calow *et al.*, 2010). In low permeability aquifers located within marginal rainfall regions, where active recharge is not guaranteed annually or even biennially, increased demand can lead to over-exploitation and failure of the aquifer during prolonged droughts.

Groundwater development has greatly increased in member states such as Malawi where borehole drilling has grown exponentially since the late 1980s. In other states, such as Namibia, where the need to conserve groundwater resources has already been recognised, the rate of groundwater development through borehole drilling has peaked and is diminishing (Figure 17.1). Although many tens of thousands of boreholes have been drilled throughout the SADC region very little in the way of usable hydrogeological data have been recorded. Nevertheless, data derived from a small number of well managed, comparatively data-rich projects located within the region can be used as district specific case studies that provide data from a variety of hydrogeological environments developed on Precambrian basement rocks, e.g. Masvingo Province in Zimbabwe (Houston, 1988; Lovell *et al.*, 1998), Mutare District in Zimbabwe (Davies, 1984), SW Tanzania (Carl Bro, Cowiconsult & Kampsax-Kruger, 1980), and south-east Malawi (Robins *et al.*, 2003).

Detailed data are inadequate with which to undertake conventional supply-and-demand analysis based on normal groundwater-resource evaluation. The groundwater resources of the main aquifers of Southern Africa are best defined by analysis of large blocks of geographically referenced hydrogeological and related data within a geo-Graphical Information System (GIS) environment. The data are available from national groundwater databases containing information obtained during the siting, drilling and testing of the many thousands of boreholes drilled in most of the SADC member states. However, much of the opportunity for data gathering has been lost in recent years.

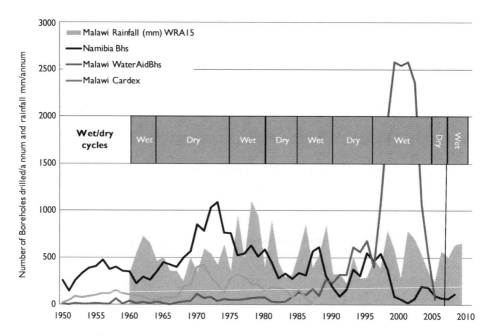

Figure 17.1 Borehole drilling statistics for Malawi and Namibia.

Just as Namibia has realised the need to conserve its resources and encourage optimal use of water so other SADC member states were rushed by the NGOs into drilling more and more boreholes. A key driver was the WHO/UNICEF Millennium Development Goals' water supply targets. An analysis of the available hydrogeological data in Malawi suggests that the sustainability of the modern drilling programmes may be short-lived and that the increased demand on the aquifer is not balanced by potential recharge. The hypothesis tested in this paper is that the explosion in drilling that has occurred in Malawi (and in other member states) in the past 20 years is not sustained by the available resource and many of the new and 'successful' boreholes are projected to go dry within a period of about 8 to 10 years.

17.2 THE MALAWI STORY

Available hydrogeological data are:

- digital topographical map coverage of Malawi, including GIS layers for elevation, drainage and infrastructure; these layers form base maps of the 17 so-called Water Resource Areas (WRAs) which are based on the main river catchments;
- the geological map of Malawi (Bloomfield, 1966) digitised to capture geological units and main structures;
- nine hydrogeological maps at 1:250 000 scale produced by the Ministry of Irrigation and Water in 1987 were scanned and the distribution of aquifers and major geological structures captured;
- other hydrogeological data from records held by the Ministry of Irrigation and Water Development, water boards, donors, NGOs, drilling companies and other agencies involved in groundwater development.

A master database of 12 000 boreholes was created from data in the Malawi borehole database and information derived from a series of small projects. The location of the boreholes was recorded in UTM co-ordinates also using information from the recent WaterAid study of water sources (WaterAid, 2010). The work focused on boreholes in Fractured Basement (FB) and Weathered Basement (WB) (Table 17.1).

Table 17.1 Attribute populations.

Attribute	Fractured basement (FB) No.	Weathered basement (WB) No.
Borehole numbers	2015	6084
Water levels	1016	4290
Water strike levels	144	376
Borehole depths	1090	4058
Borehole yields	1043	4365
Lithological logs	324	756
Specific capacities	222	1577
Transmissivities	222	1577

Statistical analysis of Specific Capacity (SC), Transmissivity (T), regolith thickness and other relevant parameters, yielded medians for each aquifer type, grouped according to each WRA. Information on some of the more critical parameters, including permeability and transmissivity and time series hydrograph data were scarce.

Assessment of potential recharge is difficult in a semi-arid to seasonal wet climate where recharge is intermittent and dependent on rare but intensive rainfall events. Monthly rainfall and Potential Evapotranspiration (PET) data were acquired for the period 1960–2009 for sub-catchments within each WRA. Subtraction of the monthly PET from monthly rainfall provided the residual potential rainfall for recharge at sub-catchment level. The monthly sub-catchment values contributed towards an average monthly rainfall rate for each WRA. Rainfall mainly takes place between December and April and averages can be derived for each annual wet season.

The rainfall available for potential recharge is greatest in the Central Rift Valley, the north-west plateau and the central plateau and least in the southern Shire valley and in the north-west central plateau. The highest occurrences of potential rainfall appear to be associated with area-specific storm/cyclonic rainfall events that occur over several adjacent catchments at any one time. Rainfall and drought fall within a broad 11-year cycle (see Figure 17.1).

Analysis of potential recharge can be made by examining the groundwater 'throughflow' or natural discharge for each aquifer type and making the critical assumption that this equates to the natural recharge. Using the Darcian approach, the groundwater gradients were derived from a piezometric contour map developed using GIS for each WRA and 'sampled' as the median gradient for each aquifer type in each WRA. Transmissivity (T) was derived with an empirical method relating T values to specific capacity for individual boreholes. Specific capacity was initially derived for each aquifer type in each WRA by extracting the final drawdown and final yield data from all boreholes that had been subjected to (short-term) testing to produce median values for each aquifer type in each WRA.

Several different approaches were used to derive transmissivity values from SC. A regional groundwater survey in southern Zambia (Baumie et al., 2007) yielded an empirical relationship between SC and T from a significant number of matching 'pairs' of data, i.e. both SC and T for individual boreholes. The equation so derived was:

$$T = 82.5SC^{1.1293} \quad [R^2 = 0.86]$$

The methodology of Betson and Robins (2003) was also applied to derive T values from SC data; the process mirrored that used in Zambia.

The slope and origin constants were applied to specific capacity data derived from the master database. These data were then analysed to produce average T values for aquifer segments in each WRA and rationalised to produce median values of transmissivity for the weathered and fractured basement aquifers within each WRA.

Flow-front (discharge) lengths were determined for the aquifer components of each WRA, i.e. from the weathered basement of the plateau area, through the fractured basement of the upper rift-valley side, down to the Quaternary fan and alluvial deposits to discharge into Lake Malawi. The flow-front is defined as the length of junction between two adjacent aquifer components within a specific WRA expressed in metres.

The following caveats need noting:

- groundwater levels on which gradients have been computed may be influenced by pumping or seasonal effects;
- SC values calculated for the master dataset rely on minimal length pumping periods and may also be influenced by depth of pump, abstraction ability and aquifer saturated thickness;
- T values have been calculated from empirical relationships derived from external data that may not necessarily be universally representative of conditions in Malawi;
- flow-front lengths have been estimated from generalised hydrogeological maps and take no account of local conditions, fracture influence and patterns or local discharge zones;
- determination of the active area for potential recharge has been derived from qualitative assessment of soil maps and assumptions that recharge will be reduced or precluded over certain zones.

Available data on water-level fluctuation and resource availability are scarce. Systematic measurements were begun in 1965 and increased incrementally to 1979. During this period climate passed from the tail end of a dry cycle (1960–1973) through the following wet cycle (1974–1979). Water-level monitoring in Malawi then collapsed following 1982 to nothing during the post-1983 period. Inspection of failed boreholes does not take place so there are no records of the occurrence of mechanical failure or of resource failure.

Data from 218 sites, where 20 or more water levels had been collected during the 1960–1985 period, were used to assess the impact of 11 year wet/dry climate cycles upon groundwater systems (Table 17.2). These water levels were collected from operational boreholes equipped with hand pumps. Water levels in low-yielding boreholes located within low-permeability formations, such as weathered and fractured basements, are markedly affected by even low rates of pumping, especially during the dry portion of an 11-year wet/dry climate cycle. There is also a tendency for over-abstraction to 'dry-out' a borehole leading to its temporary abandonment during prolonged dry periods, but aquifer water levels also show a general decline with time.

The maximum potential groundwater reserve is the volume of groundwater stored in each aquifer type under long-term conditions and is taken to represent the 'normal' condition irrespective of the possible seasonal or decadal variation due to recharge, discharge or abstraction. It represents the magnitude of the stored groundwater resources of Malawi on a WRA basis. The groundwater reserve has been computed for each aquifer type in each WRA by estimating the saturated thickness of the aquifer from the master dataset, determining by GIS the area of each aquifer and assuming the porosity of each aquifer type.

Table 17.2 Distribution of monitoring sites per WRA aquifer segment.

WRA	1	2	3	4	5	6	7	8	9	10	11	14	15	16	17
FB	4						1					2			
WB	32		1	1	5	4	7		1		3	6			

Table 17.3 Average saturated regolith thicknesses for each WRA.

WRA	1	2	3	4	5	6	7	8	9	10	11	14	15	16	17
FB	24.0	19.7	32.4	20.5			41.4	20.3	25.3	31.5	32.1	17.0	20.7	27.8	
WB	23.8	24.3	21.9	25.6	19.0	17.0	29.0	25.3	12.6	26.9	34.4	22.7	8.0	23.8	

Table 17.4 Average saturated regolith and fractured basement thicknesses for each WRA.

WRA	1	2	3	4	5	6	7	8	9	10	11	14	15	16	17
FB	32.63	40.78	37.40	20.65			41.40	34.29	45.64	38.96	38.10	28.00	20.70	38.35	
WB	33.72	37.25	24.64	30.13	24.90	37.00	27.16	37.76	31.25	28.94	38.90	30.21	22.00	35.51	

A table of 1 500 boreholes used lithological, water-level and borehole-depth data to obtain thicknesses of saturated regolith from 1 199 boreholes and regolith/fractured basement thicknesses from 1 212 boreholes (see Tables 17.3; 17.4).

It is assumed that there would be little or no recharge from areas underlain by clay soils and subsoils, bare rock, and land with high surface slopes that cause rapid runoff.

In order to arrive at an evaluation of the 'available groundwater resources' for each aquifer type in each WRA there are two potential 'resources' to consider, namely the renewable resource based upon the recharge estimation and the resource based upon the (much larger) groundwater reserve. A pragmatic approach was adopted utilising the annual renewal resource coupled with a level of 'overdraft' of the groundwater reserve, based upon the assumption that the reserve withdrawals will be replenished in the long term during periods of increased rainfall during each 11-year wet/dry cycle. The degree to which this 'overdraft' can be accommodated without excessive long-term risk is crucial to the groundwater availability. If the overdraft is not used, the available groundwater resource is greatly reduced; utilise the overdraft and available resources will be greater. In order to examine the 'fitness for purpose' of the groundwater resources of each aquifer type and WRA it is essential to recognise the spatially dispersed nature of the groundwater resources as well as the inherent ability of the aquifer to deliver particular abstraction volumes from point sources. Also to be considered is the quality of the groundwater resources in relation to ultimate use as well as the overall demand placed upon the groundwater resources.

17.3   DISCUSSION

Comparison of the derived resource-potential statistics with demand data produces some rather startling possibilities.

The vast majority of groundwater abstraction boreholes have been installed for rural water supply. Up to 1984 installation of these was the responsibility of government departments. With the advent of the World Water Decade in the

1980s, large numbers of boreholes were installed, initially by contractors under government control but subsequently by contractors under NGO control (Robins et al., 2006). In all of these circumstances only basic data were collected during siting, drilling, construction and testing of boreholes, usually enough to complete borehole-completion certificates. These data were largely collected by the drillers, few supervising hydrogeologists being available. During the period of the Mozambique civil war large numbers of boreholes were drilled for refugee supply. Many of these boreholes went unrecorded as drillers were not required to submit borehole-completion forms during that period. Hence the quality of data produced, especially post-1984, is often poor, inaccurate and lacking the required detail if recorded at all.

The recently completed WaterAid led survey (WaterAid, 2010) of all water-source points located in Malawi indicates the presence of more than 50 000 operational boreholes in Malawi and yet the database produced by the project can only vouch for 12 000 or so of these. During the WaterAid survey, water-source locations were accurately located using GPS equipment and water quality was determined using SEC meters. Some water-level data were recorded by maintenance crews during the 1948–1984 period. Unfortunately, post-1986 most boreholes were completed using minimal dimensions to minimise costs of components. Scrutiny of the rainfall data also indicates that the magnitude of rainfall, and also the long-term cyclic groundwater level changes, varies across the country from north to south. There appears to be greater variation in the south (less precipitation) than in the north, a factor which has been taken into consideration in the assessment of the possible groundwater 'overdraft' values.

The indications are that 30% of groundwater available in storage can be abstracted each year from the weathered basement formation. Taking WRA 7 by way of example, where the saturated thickness is estimated to be about 27 m, with a porosity of about 20%, then this would provide an equivalent depth of water of about 5.5 m. During year 1 of pumping when there were 226 mm of potential rainfall available then 30% abstraction would have resulted in the removal of 1.5 m of water from storage leaving 4.0 m in storage. During the next year there were 188 mm of potential rainfall, increasing storage to 4.15 m but 30% abstraction reduced the quantity in storage to 2.9 m (see Figure 17.2). With subsequent years the actual quantity pumped declines so that by 1970, after 10 years of pumping, only 0.5 m of storage remains. If only 30% of the remainder is abstracted per annum, a decline from 1.7 m/a in 1960 to 0.075 m/a by 1970, this equates with the amount of potential rainfall availability. The greater reliance upon annual recharge is shown by the graph of available water mimicking that of the annual potential rainfall. This helps explain why, during prolonged droughts, boreholes tend to 'dry up', the small quantity of storage remaining being unable to supply meaningful quantities of water as the top of any saturated zone declines to the base level of the borehole. Under such circumstances there may be some move towards borehole deepening but such efforts will, under drought conditions, only provide a much reduced yield for a short period of time. Such conditions were experienced within the Harare area during the 1991–1992 drought. Even if a porosity of 40% is applied then a similar minimal amount will be achieved just two years later. Thus boreholes can 'dry up' or fail after 10 years of use within weathered basement aquifers.

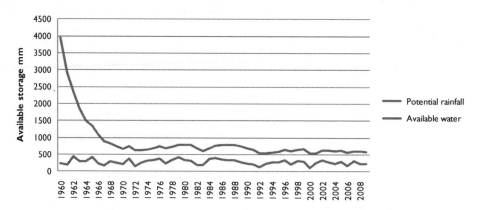

Figure 17.2 Storage available in weathered basement of WRA 7 with 20% porosity at 30% annual abstraction.

17.4   CONCLUSIONS

Despite the scarcity of data in countries such as Malawi it is possible to derive surrogate and secondary data sufficient to quantify recharge potential. Assessment of the transport mechanism in weathered basement indicates that only a small volume of lateral flow is likely, a situation exacerbated by generally low hydraulic gradients away from the escarpment. This puts pressure on abstraction sources which essentially pump from storage until their capture zone later receives an additional direct rainfall recharge.

The work in Malawi supports the hypothesis that increased drilling activity in Malawi in the past 20 years is not universally sustained by the available resource and many of the new and 'successful' boreholes are projected to go dry within a short period. Failure results from recharge being unable to sustain abstraction within the limited capture zone of an abstraction source, there being little effective lateral inflow of groundwater on the African surface plateau savannah lands which are flat and devoid of any regional hydraulic gradient. If lateral groundwater flow does occur across a source-capture zone, for example towards the rift-valley escarpment, sources are more sustainable to a degree, depending on the volume of throughflow. Where surface-water courses traverse the aquifer, albeit flowing ephemerally or largely as sub-river bed sand rivers, sustainability of borehole sources is also increased provided there is hydraulic connectivity to the aquifer. Otherwise, taking the long-term average effective rainfall (rainfall minus potential evaporation) into a weathered low permeability (0.01 m/d to 0.3 m/d) Precambrian basement aquifer (regolith and fracture zone) of average thickness 40 m with a porosity range of 20% to 40%, and mean transmissivities of 1 m²/d to 4 m²/d, the total volume of water in storage and its likely annual replenishment can be calculated.

Abstraction from the aquifer within a source capture zone may be as little as 25% of storage in isolated rural areas, more typically about 50% and as high as 75% in the more densely populated areas. Starting with a new source or group of sources and allowing abstraction at rates of 25% and 75% of total storage, coupled with effective rainfall (of the order 20% to 35% of long-term average rainfall) the water table at most borehole sources will decline year on year until the seventh or eighth year by which time pumping is no longer viable and the source has effectively failed. This is the same for the humble hand pump as it is for the larger diesel-driven sources. If demographic growth requires an increase in sources of just 1% per year, source failure will occur within a 25-year to 35-year period independently from the annual decline in available storage due to overdrawal. Amelioration and sustainability of village supplies has to resort to periodic redrilling and installation of new sources situated outside the local catchment zones of the failed boreholes.

It is likely that the same analysis can be undertaken using data derived from the Precambrian basement aquifers underlying drought-prone areas of Tanzania, Zambia, Zimbabwe, Botswana and South Africa and where groundwater occurs in similar hydrogeological environments to those found in Malawi. However, boreholes need to be inspected when they fail in order to ascertain whether it is failure due to lack of water or due to mechanical failure (or both). Such data are not currently available for much of the region. Failure does tend to happen in clusters during drought indicating the likelihood of resource failure as was observed in the 1992 drought.

REFERENCES

Baumie, R., Neukum, C., Nkhoma, J. & Silembo, O. (2007) *The groundwater resources of Southern Province, Zambia (Phase 1)*. Technical report for GTZ, BGR and Ministry of Energy and Water Development, Volume 1, Lusaka, Republic of Zambia.

Betson, M. & Robins, N. (2003) Using specific capacity to assign vulnerability to diffuse pollution in fractured aquifers in Scotland. In: Krasny, J. & Sharp, J.M. (eds.) *Groundwater in Fractured Rocks: Selected Papers from the Groundwater in Fractured Rocks International Symposium, Prague, 2003*. UK, Taylor and Francis. Chapter 33, pp. 495–505.

Bloomfield, K. (1966) *Geological Map of Malawi*. London, UK, Edward Stanford.

Calow, R.C., MacDonald, A.M., Nicol, A.L. & Robins, N.S. (2010) Ground water security and drought in Africa: Linking availability, access and demand. *Ground Water* 48, 246–256.

Carl Bro, Cowiconsult & Kampsax-Kruger (1980) *Water master plans for Iringa, Ravuma and Mbeya regions, the geomorphological approach to the hydrogeology of the basement complex*. Report Carl Bro, Cowiconsult & Kampsax-Kruger, report for Danish International Development Agency. [Online] Available from: www.sadcgwarchive.net [Last Accessed 30 July 2012].

Davies, J. (1984) *The hydrogeology of the Nyamazura area*. Wallingford, UK, British Geological Survey. Internal report WD/OS/84/16.

Houston, J.F.T. (1988) Rainfall runoff recharge relationships in the basement rocks of Zimbabwe. In: Simmers, I. (ed.) *Estimation of natural groundwater recharge*. Dordrecht, Holland, Riedel. pp. 249–265.

Lovell, C.J., Simmonds, L., Waughray, D.K., Semple, A.J., Mazhangara, E., Murata, M.W., Brown, M., Thompson, D.M., Chilton, P.J., Macdonald, D.M.J., Conyers, D., Butterworth, J.A., Mugweni, O., Moriarty, P., Bromley, J., Batchelor, C.H., Mharapara, I., Mugabe, F.T., Mtetwa, G. & Dube, T. (1998) *The Romwe catchment study, Zimbabwe: The effects of changing rainfall and land use on recharge to crystalline basement aquifers, and the implications for rural water supply and small scale irrigation*. Wallingford, UK, Institute of Hydrology. Technical report.

Robins, N.S., Davies, J., Hankin, P. & Sauer, D. (2003) Groundwater and data – An African experience. *Waterlines* 21, 19–21.

Robins, N.S., Davies, J., Farr, J.L. & Calow, R.C. (2006) The changing role of hydrogeology in semi-arid southern and eastern Africa. *Hydrogeology Journal* 14, 1483–1492.

Smith-Carrington, A.K. & Chilton, P.J. (1983) Groundwater resources of Malawi. Republic of Malawi, Department of Lands, Valuation and Water. Unpublished report, Institute of Geological Sciences, UK.

Titus, R., Friese, A. & Adams, S. (2009) A tectonic and geomorphic framework for the development of basement aquifers in Namaqualand – A review. In: Titus, R., Beekman, H., Adams, S. & Strachan, L. (eds.) *The basement aquifers of Southern Africa*. Pretoria, South Africa, Water Research Commission. pp. 5–18. WRC Report No. TT pp. 428-09.

WaterAid (2010) Spreadsheet water point mapping tool. [Online] Available from: http://www.wateraid.org/mapper [Last Accessed June 14, 2011].

Chapter 18

Preliminary assessment of water-supply availability with regard to potential shale-gas development in the Karoo region of South Africa

P.D. Vermeulen
Institute for Groundwater Studies, University of the Free State, Bloemfontein, South Africa

ABSTRACT

South Africa is a country blessed with minerals, but without oil and conventional gas resources and with a shortage of energy, resulting in an energy import. During the period 1965 to 1975 Soekor (Pty) Ltd explored for oil and gas in the Karoo, a pristine semi-desert area in South Africa. Gas was found in the tight shale formations of the Ecca Supergroup, between 2 500 and 4 000 m below the surface. It has been calculated that the deposit may be as large as 35.3 TCM (or 485 Tcf as is the custom in the oil and gas industry to refer to the volume), and that the worst case scenario will be 1.1 TCM (which makes it the fifth largest deposit in the world). Exploration companies applied for exploration licences, but due to an uproar by environmental groups a moratorium was placed on any activities. The biggest issue, apart from environmental negatives, was the shortage of water resources. Is it possible to harvest enough water in the Karoo aquifers for this process?

This paper will deal with the process of fracking (hydro-fracturing) and the different views in South Africa regarding the availability of water, how to deal with the dolerite and the other unknowns and the proposed way forward. The problem is that the unique geology (presence of dolerite dykes and sills) of South Africa makes case studies from other parts of the world worthless to a degree.

Note: No scientific work has been done in South Africa regarding the groundwater issues during hydro-fracturing as yet, and most statements are based on observations made by the author during two study tours to different sites in the USA and interviews with relevant role-players. (TCM = Trillion Cubic Meter and TcF = Trillion cubic Feet)

18.1 INTRODUCTION TO SHALE-GAS EXPLOITATION

Historically gas has been extracted from permeable formations from which gas can flow. Vertical wells were drilled to the target formation, from where gas flows into the well. Typically the target formations comprised sandstone which has undergone some secondary processes to enhance the properties of the rock which will allow the gas to flow.

Shale may also contain gas, but has a poor ability to transmit fluids. As a result, gas cannot readily migrate in shale formations. Fracking is a well-stimulation technique aimed at opening existing cracks or creating new ones to more effectively and efficiently extract gas from low-permeability (or tight) shale formations. It entails drilling vertical wells until the target horizon is encountered, whereafter horizontal wells are drilled (Figure 18.1). Using water, sand and chemicals, sections of the horizontal well are systematically subjected to very high pressures (as high as 70 000 kPa, or 700 bar) to open existing fractures or create new ones. Fractures are generally less than 1 mm wide and the sand is used as a proppant to keep the fractures open. Typically the section of well fracked at any one time is between 100 m and 150 m in length, while the induced fracture zone extends between 200 m and 300 m from the horizontal well. The drilling of a single well takes about 1 month to complete and the fracking of the well 14 d in total.

Fracking has been used to stimulate oil, gas and water wells since 1947, when first used at Hugoton. However, what has changed significantly in the past few years is the simultaneous use of four technologies, namely:

- directional drilling;
- high volume fracking-fluids;
- slick-water additives;
- multi-well pads.

High-volume fracking has greatly increased the ability to extract natural gas from tight rock. Shale beds are often less than 100 m thick, so fracking wells are usually drilled horizontally to extend through as much of the bed lengthwise as possible. The horizontal component of the well can be in the order of 1.5 km to 3 km in length.

The wells are sealed with four casings within one another. High-pressure water, sometimes as high as 70 000 kPa, is then sent down this pipe. This system acts like a hydraulic ram where great force can be applied over a large area by introducing high pressure into the ram from a small entry point. This force splits the shale, creating the numerous small fractures. Proppants are then forced into the fractures to keep the fractures open.

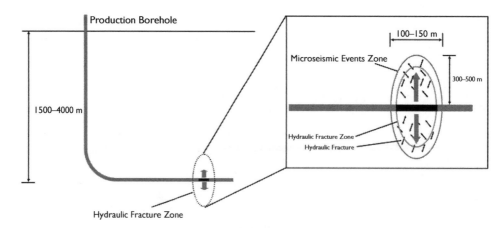

Figure 18.1 Hydraulic fracturing process producing hydraulic fracture zones perpendicular to borehole.

The fluid injected into the rock typically comprises a slurry of water, proppants and chemical additives. Additionally, gels, foams, and compressed gases, including nitrogen, carbon dioxide and air can be injected. Types of proppant include silica sand, resin-coated sand and man-made ceramics. Chemical additives are selected to suit the specific geological situation, protect the well and improve its operation. Typically the fracking fluid comprises 90% water, 9.5% proppant and 0.5% chemicals. The chemicals may consist of up to 13 classes of additives (proppant excluded). The composition of the fracking fluid used may vary from one geological basin or formation to another or from one area to another in order to meet the specific needs of each operation, but the range of additive types available for potential use remains the same. There are a number of different products for each additive type; however, only one product of each type is typically used in any given fracking job. The selection may be driven by the formation and potential interactions between additives. Additionally, not all additive types will be used in every fracking job.

The total amount of fracking additives and water used in the fracking of horizontal wells is considerably greater than for vertical wells. This suggests that the potential environmental consequences could be proportionally larger for horizontal well-drilling and fracking.

18.1.1 Water required for high-volume fracking

According to Dr Jim Richenderfer (Technical Programs Director: Susquehanna River Basin Commission, which regulates water use for fracking sites in Pennsylvania) between 14 000 m^3 and 16 000 m^3 freshwater is used in the drilling and fracking of one horizontal well. This range was based on water supplied to 724 fracked wells. This is supplemented with flowback water from other boreholes. In other areas, water use can be as low as 10 000 m^3 while in New York State estimated use as high as 26 000 m^3 has been recorded (SGEIS Report, 2011). For the sake of this review, it is accepted 20 000 m^3 of water will be required to frack each well in the Karoo.

Given the claims of excessive water requirements for fracking and the impact on the environment, this volume of water needs to be put into perspective:

* about 5 000 m^3 is required to irrigate 1ha of crop in the Karoo, thus the volume of water needed to frack one well is equivalent to that used to irrigate 4 ha of land. The key difference in water demand is that the irrigated water is applied over 3 months to 4 months while that used by fracking is used over a period of 5 d;
* the average daily water consumption of a Karoo town (e.g., Beaufort West) is 8 500 m^3, thus the volume of water needed to frack one well is equivalent to that used over 2.5 d by the town. However, the rate of water required for fracking is about half of that of Beaufort West, and is required for a period of 5 d;
* if the water were to be trucked to site, it would require about 650 tanker loads to get the required volume of water to site. According to calculations from the SGEIS Report (2011) it is estimated that 1 148 heavy truck loads would be made in total to each well site during drilling and fracking for the first borehole drilled and then a further 843 heavy truck loads for all other boreholes drilled and fracked at one site.

18.1.2 Well-pad density

One issue for residents and visitors in the Marcellus Shale region in the USA is the density of well pads in their area. Many people want to know how frequently they can expect to see drilling facilities in a particular area. As no maps are available that provide locations of drilling in Pennsylvania, figures showing pad density are generally estimates.

Regulatory bodies in the USA allow a maximum density of 1 well pad per 260 ha. In recent shale-gas operations in Canada, multi-well pads are about 5 km apart. A similar spacing would be pursued by companies in South Africa if hydrocarbons are discovered and developed in the Karoo. According to Mr Jan-Willem Eggink, (Upstream General Manager, Shell SA) in 2011, these sites will be typically 4 km to 5 km apart in the Karoo. Only a limited part of the Karoo will be accessible for shale-gas drilling, due to quite a number of exclusions, e.g. topographical constraints, SKA, SALT, townships, nature parks, etc. It is estimated that some 28% of their area of 90 000 km^2 is accessible. They cannot develop the entire 28% since they need to find the 'sweet-spots' in this area (areas without dolerite intrusions). From experience in the US this is again only a small proportion, roughly between 15% and 30%. So in reality only 5% to 10% of the area may be available for commercially attractive activities, i.e. some 4 500 km^2 to 9 000 km^2 in the area they are applying for. So if we assume 25 km^2 per well pad, then there will be 180 to 360 well-pads. Each well pad may have some 30 wells, i.e. some 5 000 to 10 000 wells. If we assume 5 km^2 per pad, it will be up to 1 500 pads (based on 6 to 8 wells per pad).

18.1.3 Volumes of flow-back water

Once fracking procedures have been completed and pressure is released, the direction of fluid flow reverses, and fluids flow toward the well. The well is developed (or cleaned) by allowing water and excess proppant to flow up through the well to the surface. This is referred to as flow-back water.

Flow-back water recoveries reported from horizontal wells in the Marcellus Shale in northern Pennsylvania range between 9% and 35% of the fracking fluid pumped into the well, with an average of 12% (Richenderfer, 2011). This volume is generally recovered within 2 weeks to 8 weeks, whereafter the rate of water production sharply declines and levels off at approximately a barrel per day for the remainder of its producing life. This is called production water. Limited time-series data indicate that more than 60% of the total flow-back occurs in the first 4 d after fracking (sometimes as much as 95% returns in the first 5 d).

High Total Dissolved Solids (TDS), chlorides, surfactants, gelling agents and metals are the components of greatest concern in spent gel and foam fracking fluids (i.e., flow-back). Most fracking fluid additives used in a well can be expected in the flow-back water, although some are expected to be consumed in the well (e.g., strong acids) or react during the fracking process to form different products (e.g., polymer precursors).

The Marcellus Shale was deposited under marine conditions (as were sediments of the Ecca Group in South Africa) and contains high concentrations of the following (URS, 2009):

* dissolved solids (chlorides, sulphates, and calcium);
* metals (calcium, magnesium, barium, strontium);

- suspended solids;
- mineral scales (calcium carbonate and barium sulphate);
- bacteria – acid-producing bacteria and sulphate-reducing bacteria;
- friction reducers;
- iron solids (iron oxide and iron sulphide);
- dispersed clay fines, colloids and silts;
- acid gases (carbon dioxide, hydrogen sulphide).

Given the similar environments of deposition, long residence time and the lack of circulation, similar groundwater qualities can be expected from the deep Karoo Supergroup sediments.

Concern has been expressed regarding the mobilisation of radioactive elements such as uranium and other heavy metals during the fracking process and their emergence at the surface with the fracking waters. Studies done on the fine-grained sedimentary rocks of the Karoo Supergroup (and especially on the formations under discussion) indicate that the shale is not enriched with any of the elements of concern (Cole, 2008). These geological units were deposited under different conditions to those of the shale being targeted for gas exploitation. Further, uranium is mobilised by oxidising groundwater while the groundwater at depth is in a reducing state. Nonetheless it is acknowledged that little is known about groundwater quality beyond a depth of 300 m, and the concentrations of heavy metals and radioactive elements at depth are unknown. However, work currently done by the IGS under supervision of the author, indicates the presence of radio-active material in the shale of the Ecca formations in the Waterberg Coalfield.

Operators reuse flow-back water for subsequent fracking operations at the same multi-well pad or those adjacent to it. Reuse involves either mixing the flow-back water with freshwater or treatment options prior to reuse. Flow-back water is contained in sealed tankers (Figure 18.2).

The mixing option involves blending flow-back water with freshwater to make it usable for future fracking operations. Because high concentrations of various constituents in flow-back water may adversely affect the desired fracking-fluid properties, reuse without blending or treatment is not possible. Concentrations of chlorides, calcium, magnesium, barium, carbonates, sulphates, solids and microbes in flow-back water may be too high to use as is, meaning that some form of physical and/or chemical separation is typically needed prior to recycling flow-back. In addition, the practice of blending flow-back with freshwater involves balancing the additional freshwater needs with the additional additive needs.

After a number of reuse cycles, the water is transported to treatment plants. These are reverse osmosis techniques (if the water quality allows it), or thermal distillation. Thermal distillation uses evaporation and crystallisation techniques that integrate a multi-effect distillation column, and this technology may be used to treat flow-back water with a large range of parameter concentrations. For example, thermal distillation may be able to treat TDS concentrations from 5 000 mg/l to 150 000 mg/l, and produce water with TDS concentrations of between 50 mg/l to 150 mg/l. The resulting residual salt would need appropriate disposal. This technology is resilient to fouling and scaling, but is energy-intensive and has a large footprint.

Figure 18.2 Tankers used to store flow-back water until treatment and disposal.

18.1.3.1 Shale gas in South Africa

Much of the area of the Republic of South Africa comprises rocks of the Karoo Supergroup. This geological unit has been described by Truswell (1977), Catuneanu *et al.* (2005), and others.

Carbon-rich units of the Ecca Group of the Karoo Supergroup are the target zones for shale-gas exploration, principally the Whitehill and Collingham Formations. These units need to be encountered where pressure and temperature conditions are favourable for gas generation. The depth and thickness of the target zones is relatively well known, being informed by geophysical exploration and drilling of 24 deep wells by Soekor (Pty) Ltd[1] in the 1960s and 1970s (Roswell and de Swardt, 1976). The basal Prince Albert Formation and the upper Waterford and Fort Brown Formations are also of interest.

Little is known about the prevalence and orientation of dolerite at depths greater than that typically reached by boreholes. The Karoo represents an erosion surface from which thousands of metres of sediment have been eroded and removed. What is now seen at the surface used to be deep below the surface. By extension, it is understood that the dolerite at depth has a similar character and morphology to that currently seen at the surface.

As far as is known, South Africa is the only known instance where shale-gas targets have been intruded by dolerite. This makes the South African situation unique; the

---

1 A new State oil company, PetroSA, was formed in January 2002 from the merger of three previous entities Mossgas (Pty) Limited, Soekor (Pty) Limited, and parts of the Strategic Fuel Fund Association.

ready extrapolation of knowledge from elsewhere in the world to the South African situation should therefore be done with caution.

18.1.4   Hydrocarbon exploration

Exploration of the Karoo Basin for oil was undertaken in the 1960s, with some gas reserves being detected (Roswell and de Swardt, 1976). Gas flowed from well Cranemere 1/68 from a depth of 3 600 m at a rate of 51000 million cubic meter (MCM) per day for a short period. However, the resources were considered unviable at the time and no further exploration has been undertaken in the past 40 years.

Geological units with a total organic carbon of between 3% and 12% are being targeted for shale-gas exploration. If compared to the Marcellus Shale (0.3% to 20% carbon and a thickness of 12 m to 270 m) and the Barnett shale (0.5% to 13% carbon and a thickness of 15 m to 300 m) in the USA, the Whitehill formation (0.5% to 14.7% carbon and a thickness of 1 m to 72 m) and the Prince Albert formation (0.35% to 12.4% carbon and a thickness of 30 m to 420 m) in the Karoo, obtained from the Soekor (Pty) Ltd data, compare favourably.

There are currently a number of companies that have exploration rights to investigate natural gas resources in Karoo-type formations (Figure 18.3).

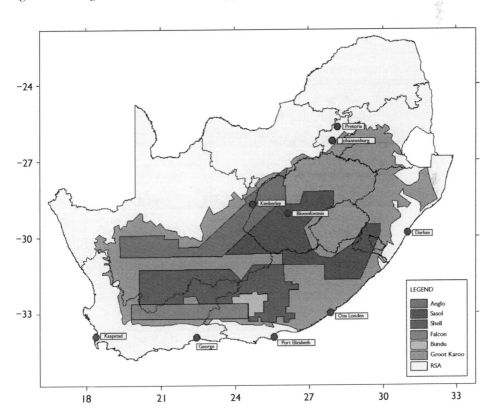

Figure 18.3 Regional map of South Africa showing the original exploration rights and companies associated with these permits.

18.1.5 Geohydrology

Karoo aquifers are secondary in character, and owe their storage and transmission capabilities to weathering, fracturing, faulting and the intrusion of dolerite bodies. These secondary aquifers are both heterogeneous and anisotropic, resulting in properties varying significantly over short distances. A dual porosity model is considered appropriate for these systems, where most groundwater is stored in the matrix and the transmission of groundwater takes place in secondary openings such as joints and fractures.

18.1.6 Possible sources of water

Much of the Karoo experiences rainfall ranging between 200 mm/a to 400 mm/a. Vegetation in the Karoo reflects the low rainfall of the region and together with characteristic dry river beds and infrequent occurrence of dams, a visual image is created of a region without water. The availability of sufficient water for fracking is a key consideration when assessing whether or not fracking will be viable in the Karoo, and a number of possible sources of water exist:

- development of local groundwater supplies – this includes exploiting the 4 000 breccia plugs in the western Karoo, and the hydraulic fracturing of the aquifers of the Karoo;
- trucking in or piping surface water from elsewhere – either by road, rail, pipeline or a combination thereof. This will put an additional burden on the roads and infrastructure. Transfer of water, either freshwater or seawater, clearly has a number of limitations and must be subject to a proper study;
- piping seawater or desalinated seawater from the coast – this will only be feasible if the water is purified and piped across the escarpment, which will be a very costly exercise;
- water from the Orange River – the excess water is already being allocated to previously disadvantaged farmers;
- Mr Jan-Willem Eggink, (Upstream General Manager, Shell SA) has stated publicly that they will not compete with water users in the Karoo for potable water. They consider exploiting the deeper saline aquifers for water. A few hundred metres below the surface lie under-utilised aquifers. Systematic exploration and research indicates that the harvest potential of these aquifers ranges from 5 000 $m^3/km^2/a$ to in excess of 25 000 $m^3/km^2/a$ (Baron *et al.*, 1998).

It is noteworthy that hydrogeological investigations of Karoo aquifers are generally restricted to the upper 100 m. Modern water-drilling equipment has increased the depth of drilling to 150 m, but few boreholes are drilled deeper than 300 m. Data extracted from the National Groundwater Database in 2008 and presented by Golder Associates Africa (Pty) Ltd (2011) indicated only 6% of the 2 323 boreholes were drilled deeper than 100 m.

Woodford and Chevallier (2002) reported similar findings from analysing 67 borehole logs totalling 13 799 m (average depth of 206 m) in the vicinity of Victoria West. They found that 73% of all water interceptions occurred within 80 m of the surface.

Kent (1949) assessed thermal springs in South Africa and reported that 12 of the 74 thermal springs are associated with rocks of the Karoo Supergroup. Of these, 8 are located in the target areas for shale gas and are classified as warm springs. Most rise alongside dolerite dykes from depths of about 600 m to 1 300 m. While the thermal springs are of relatively shallow origin, they do suggest that groundwater at greater depths is more saline than that in the upper 150 m.

Given that unconventional shale-gas exploration will target geological formations of between 3 000 m and 5 000 m below ground, a significant barrier exists between the aquifers of the Karoo and the target zones for shale gas. Given the low permeabilities reported by Roswell and De Swardt (1976), and accepted thickness of aquifers in the Karoo, it is conceptually difficult to accept any connection between the two bodies. However, the integrity of this barrier could be compromised by leaking casings. It should be noted that one school of thought in South Africa believes that the Karoo basin is under pressure, and that the potable water will be severely compromised after production of the wells has ceased and the wells are sealed.

18.1.7 Volume of groundwater in the Karoo formations of SA

Van der Voort (2001) determined the porosity values of shallow fractured sedimentary rocks (less than 100 m below the surface) from geological cores of different formations in the Karoo and obtained values of between 3% and 15% with an average of 8%. This implies that each cubic metre of rock contains on average 0.08 m^3 of water (or 80 l). A hectare of land thus contains in the order of 80 m^3 of water for each 1 metre of saturated thickness. If a saturated thickness of 80 m is taken (i.e., from an average water level of 20 m below the surface to 100 m), an amount of 6 400 m^3 is stored in 1 ha. The groundwater recharge ranges between 1% and 5% of the annual rainfall with an average of 3% and for an annual rainfall of 480 mm, the annual recharge will be about 144 m^3/ha or 2.25% of the total amount of groundwater stored in the first 100 m of the formation.

The porosity of sandstones and of shales can be used as a rough measure of the original thickness of overburden. Maxwell (1964) shows that the porosity of clean quartzitic sandstones decreases as a linear function of increasing burial and also to some extent with increasing age. In areas with a high thermal gradient the reduction in porosity is more rapid than in those with a low gradient, i.e. it is partly temperature-dependent. According to Maxwell's studies sandstones can be expected to have a porosity of about 10% to 15% at a depth of 3 000 m and about 5% at a depth of 5 000 m. In the sandstones studied by him the processes chiefly responsible for porosity reduction were compaction and cementation. Unlike the porosity of 'clean' sandstones, the porosity of shales decreases very rapidly in the first few hundred metres of burial as interstitial water is expelled and more slowly thereafter. Maxwell (1964) shows that shale porosity is between 5% and 10% at a depth of about 2 000 m, and 5% at slightly over 3 000 m. According to Rumeau and Sourisse (1972), however, shale porosity may be strongly influenced by the age of the sediments (duration of burial). Thus at 3 000 m it may be between 15 to 20 percent at 2 000 m in very old (Paleozoic) sediments. It seems highly probable that shale porosity is also influenced by temperature.

From a depth of 100 m to 3 000 m below the surface the only laboratory porosity estimations are those obtained from the 24 Soekor (Pty) Ltd boreholes drilled in the 1960s. The porosity estimates vary between 0.1% and 8% with an average of 2.5%. It is thus expected that the Karoo formations contain water at large depths but that the quality of the water is not good (for domestic purposes). The permeability (hydraulic conductivity) is extremely low (less than 10 m/d to 11 m/d) and hydraulic fracturing is thus required for the water to flow freely from the rock.

The shale-gas exploration companies can thus use hydro-fracturing in the shallower Karoo formations to increase the permeability of the rock and these deeper groundwater resources (e.g., from 300 m to 700 m) could be used as the water for the fracking of shale gas.

If the companies are going to use deep groundwater to satisfy their need for fracking, the question will be what the impact of such operations will be on the shallower fresh groundwater aquifer. From practical experience in the Karoo (Mooikraal Underground Coal Colliery), we know that a shallow water table still exists above mined-out underground collieries which are deeper than 100 m below the surface. During the fracking process for water at shallow depths, no harmful chemicals will be used (only water and sand), which will make it a safe, feasible option.

Proper detailed investigation is required to assess the viability of supplying water for fracking in the Karoo. Appropriate exploration, testing and management of groundwater resources will be required if groundwater resources are to be used. Targets for exploration include dolerite-contact zones, breccia plugs and localised fracturing.

It is advisable to drill a few test well fields in order to determine if it is viable to extract enough water from the Karoo aquifers without depleting them. If successful, this will put stakeholders at ease, as groundwater is the lifeline of the Karoo. The key to the successful and sustainable use of any resource – including groundwater – is proper development and management. In the Denton area in Texas a number of boreholes are drilled on a farm long before fracking commences, and low abstraction over a period of time, pumped into a constructed dam, ensures that the groundwater is not depleted. The water is then piped with coupled irrigation pipes to the well site. This practice also decreases the volume of traffic in the area.

18.1.8 Positives and negatives regarding fracking in the Karoo

Positives:

- in a country that is a net importer of energy, shale gas can improve the deficit if the reserves are proven to be substantial;
- techniques have improved markedly in the past few years. More casings are used down the hole, more environmentally friendly chemicals are used and modern treatment facilities are built to remove the salts from the flow-back water. Enclosed containers are used for the flow-back water and all operations are done on geomembranes.

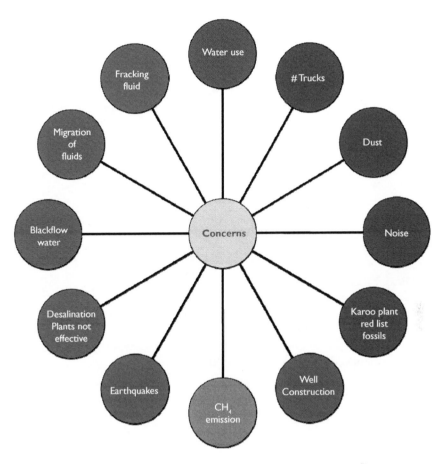

Figure 18.4 Level I concerns.

Negatives:

The negatives and concerns are listed below in Figure 18.4. This is the personal opinion of the author and may differ from what other people may list as important. There are always risks associated with any mining operation and one can assume with confidence that the more operations (mining) taking place, the higher the risks.

The main concerns are the availability of water, the number of trucks on the roads (and the integrity of the roads), the visible effect that it will have on the pristine landscape, the effect of the dolerite dykes and well construction and integrity.

18.2 CONCLUSIONS

Shale gas plays an important part in the energy needs of the USA (and also elsewhere), and this may be the same in South Africa. However, there are still a huge number of issues that needs to be resolved before such operations can be allowed in the Karoo.

Do we have the gas resources that are being predicted, will there be enough water in the Karoo (or will water be transported from somewhere else), and can the Karoo be rehabilitated after such an onslaught of activities? The answer is still not clear; a huge research effort is still required to answer these questions.

Since this paper was presented, the moratorium was lifted on exploration of shale gas in South Africa. The author is of the opinion that the Groundwater Division of the Geological Society of South Africa should take a unified formal stance on the shale-gas issue regarding a number of matters. These include:

- What type of research must be done, and who must do it?
- Who should fund such research? Should it be government, or should industry fund it?
- Who should review the work in order to give credibility to the work? This is important to ensure that it will not just be regarded as work paid for by industry (as was the case with some of the studies that have already been done).

ACKNOWLEDGEMENTS

The researchers at the Institute for Groundwater Studies for their input. The USGS in Pennsylvania for organising a tour through the Marcellus Shale area, the mining companies for allowing us on site and the Pennsylvania State Protection Agency for valuable information during meetings with them.

REFERENCES

Baron, J., Seward, P. & Seymour A. (1998) *The groundwater harvest potential map of the Republic of South Africa*. Pretoria, South Africa, Directorate of Geohydrology, Department of Water Affairs and Forestry. Technical report Gh 3917.

Catuneanu, O., Wopfner, H., Eriksson, P.G., Caincross, B., Rubidge, B.S., Smith, R.M.H. & Hancox, P.J. (2005) The Karoo Basin of south-central Africa. *Journal of African Earth Sciences* 43, 211–253.

Cole, D.I. (2009) A review of uranium deposits in the Karoo Supergroup of South Africa. In: *Proceedings of the International AAPG Conference, Cape Town, South Africa*, October pp. 26–29, 2008.

Eggink, J. Upstream General Manager, Shell SA. (Personal communication, March, May, June 2011).

Golder Associates Africa (2011) *Groundwater report in support of the EMP for the South Western Karoo Basin gas exploration application project – Western Precinct*. Midrand, South Africa, Golder Associates Africa (Pty) Ltd.

Kent, L.E. (1949) The thermal waters of the Union of South Africa and South West Africa. *Transactions of the Geological Society of South Africa* 52, 231–264.

Kuuskraa, V., Stevens, S., Van Leeuwen, T. & Moodhe, K. (2011) World shale gas resources: An initial assessment. Prepared for: United States Energy Information Administration, 2011. World shale gas resources: An initial assessment of 14 regions outside the United States.

Maxwell, J.C. (1964) Influence of depth; temperature and geological age on porosity of quartzose sandstone. *Bulletin American Association of Petroleum Geology* 48, 697–709.

Richenderfer, J. Technical Programs Director: Susquehanna River Basin Commission, Harrisburg, Pennsylvania, USA. (Personal communication, 2011).

Roswell, D.M. & de Swardt, A.M.J. (1976) Diagenesis in Cape and Karoo sediments, South Africa and it's bearing on their hydrocarbon potential. *Transactions Geological Society of South Africa* 79, 181–245.

Rumeau, L. & Sourisse, C. (1972) Compaction, diagenese et migration dans les sediments argileux. *Bulletin of the Centre de Recherches de Pau-S.N.P.A.* 6, 313–345.

SGEIS (2011) Supplemental generic environmental impact statement on the oil, gas and solution mining regulatory program; Revised draft, New York, NY, New York State Department of Environmental Conservation.

Truswell, J.F. (1977) *The Geological Evolution of South Africa*. Cape Town, South Africa, Purnell.

URS Corporation (2009) Water-related issues associated with gas production in the Marcellus Shale. Report prepared for NYSERDA, Albany, NY. In: Alpha Environmental Consultants (2009) Technical consulting reports prepared in support of the draft supplemental generic environmental impact statement for natural gas production in New York State, Chapter 2.

Van der Voort, I. (2001) Risk based decision tool for managing and protecting groundwater resources, PhD thesis, University of the Free State, Bloemfontein, South Africa.

Woodford, A.C. & Chevallier, L. (eds.) (2002) *Hydrogeology of the main Karoo Basin current knowledge and future research needs*. Pretoria, South Africa, Water Research Commission. WRC Report No TT 179/02.

Chapter 19

Compiling the South African Development Community Hydrogeological Map and Atlas: Opportunities for the future[†]

K. Pietersen[1], N. Kellgren[2] & M. Roos[3]

[1]SLR Consulting, Faerie Glen, South Africa
[2]Sweco International, Gothenburg, Sweden
[3]Council for Geoscience, Pretoria, South Africa

ABSTRACT

The Southern African Development Community (SADC)[1] Hydrogeological Map and Atlas (HGM) project produced a comprehensive, interactive web-based general hydrogeological map at a scale of 1: 2 500 000. The SADC-HGM provides information on the extent and geometry of regional aquifer systems, and primarily serves as a base map for hydrogeologists and water-resource planners. The hydro-lithology base map was compiled from the SADC geology map prepared by the South African Council for Geoscience. This was done through linking the stratigraphy to the rock types which simplified the 12 hydro-lithological classes that had been mapped. The identified hydro-lithologies were grouped into aquifer types: A. Unconsolidated intergranular aquifers; B. Fissured aquifers; C. Karst aquifers; and D. Low-permeability formations. The aquifer types were grouped into eight classes according to aquifer productivity. The step of assigning aquifer productivity was a lengthy process where all available geological and hydrogeological materials were considered for each area (in map terms, a polygon or a set of polygons). In effect, the rock-type polygons were overlain, using a GIS or manually, with relevant reference layers (primarily the scanned national hydrogeological maps), and an essentially 'manual', expertise-based decision was made for each area. This was followed by national contact persons from SADC Member States to verify and update pertinent data to improve the hydrogeological map, reassigning production classes according to improved knowledge, or even re-defining them if necessary. The SADC Hydrogeological Mapping Project has delineated 14 transboundary aquifer systems on the basis of inferred continuous and transmissive aquifers, SADC hydro-lithological boundaries, and sub-basin river boundaries. The map should be used to guide policy-making and to influence political decision-making on water-resource issues and assist transboundary groundwater planning and management by water-resource planners.

[†] Parts of this paper were originally presented at the International Conference 'Transboundary Aquifers: Challenges and New Directions' (ISARM2010).

[1] The Southern African Development Community (SADC) is an inter-governmental organisation with a membership of 15 Member States, namely; Angola, Botswana, Democratic Republic of Congo (DRC), Lesotho, Madagascar, Malawi, Mauritius, Mozambique, Namibia, Seychelles, South Africa, Swaziland, United Republic of Tanzania, Zambia and Zimbabwe.

19.1 INTRODUCTION

The Southern African Development Community Hydrogeological Map and Atlas (SADC-HGM) was compiled during June 2009 to March 2010 with the objectives of improved:

- understanding of groundwater occurrence in the SADC region;
- cooperation between member states;
- understanding of groundwater resource management.

The project outputs include:

- an interactive web-based hydrogeological map of the SADC region;
- institutional capacity for hydrogeological map production;
- use of hydrogeological maps for groundwater resource evaluation.

19.2 BACKGROUND

Hydrogeological maps indicate groundwater occurrence through illustration of aquifer system extent, geometry, water levels, hydraulic parameters, and flow and hydrochemical characteristics. The preparation of a hydrogeological map of the SADC region was identified as a priority in the Regional Strategic Action Plan for Integrated Water Resources Development and Management for 2005–2010 (SADC, 2005) (Box 1 and

---

Box 1: Regional Strategic Action Plan on Integrated Water Resources Development and Management (RSAP-IWRM). Annotated Strategic Action Plan 2005–2010 (SADC, 2005)

The mission of the RSAP-IWRM is *'to provide a sustainable enabling environment, leadership and coordination in water resources strategic planning, use and infrastructure development through the application of integrated water resource management at Member State, regional, river basin and community level'*.

Four strategic areas are identified:

- Regional Water Resources Planning and Management
- Infrastructure Development Support
- Water Governance
- Capacity Building.

A number of projects are distributed within the strategic areas. For example, the area focussing on Regional Water Resource Planning and Management has the following projects:

- RWR 1: Consolidation and Expansion of SADC HYCOS
- RWR 2: Standard Assessment of Water Resources
- **RWR 3: Groundwater Management Programme in SADC**
- RWR 4: Support for Strategic and Integrated Water Resources Planning
- RWR 5: Dam Safety, Synchronisation and Emergency Operations.

Box 2: Groundwater Management Programme in SADC (ten projects listed in order of priority)
(SADC, 2005)

1. Capacity Building within the Context of Regional Groundwater Management Programme
2. Develop Minimum Common Standards for Groundwater Development in the SADC Region
3. Development of a Regional Groundwater Information System
4. Establishment of a Regional Groundwater Monitoring Network
5. **Compilation of a regional Hydrogeological Map and Atlas for the SADC Region**
6. Establish a Regional Groundwater Research Institute/Commission
7. Construct a Website on Internet and publish quarterly News letters
8. Regional Groundwater Resource Assessment of Karoo Aquifers
9. Regional Groundwater Resource Assessment of Precambrian Basement Aquifers
10. Groundwater Resource Assessment of Limpopo/Save Basin.

Box 2). The rationale was provision of a synoptic overview of the hydrogeology of the SADC region to enhance groundwater-resource management.

The compilation of the SADC-HGM was enabled by publication of the SADC Geological Map (Hartzer, 2009) by the Council for Geoscience of South Africa at a scale of 1: 2 500 000. Production of the SADC-HGM made use of the regional lithostratigraphy developed for the SADC geology map. These lithostratigraphic boundaries were used in simplified form to define the extent of aquifer types, these units being divided upon borehole yield potential.

Although no reliable groundwater-use statistics are available, it is understood that significant groundwater resources remain untapped within the SADC region that could be sustainably developed for use in:

- rural water supply;
- urban water supply;
- water security;
- food security;
- environmental services.

By identifying the location of these groundwater resources the SADC Hydrogeological Map and Atlas provide the necessary information to support sustainable groundwater-resource development at national and at regional transnational levels.

19.2.1   The science (and art) of hydrogeology mapping

Struckmeier and Margat (1995) reviewed national and international hydrogeological mapping methodologies to produce a guide for the compilation of hydrogeological maps that has been adopted as an international standard by UNESCO. Hydrogeological maps primarily aim to show the inter-relationship of groundwater within geological formations dependent upon the degree of information and data available. Given the disparity of data distribution within the SADC nations two types are commonly used in the SADC region (Struckmeier and Margat, 1995):

• general hydrogeological maps and groundwater-system maps associated with reconnaissance or scientific levels are suitable tools to introduce the importance of water (including groundwater) resources into the political and social development process;
• parameter maps and special purpose maps are part of the basis of economic development for planning, engineering and management; they differ greatly in content and representation according to their specific purpose. An example of a special purpose map would be one which shows areas of highly protected groundwater resources, used for waste-disposal planning.

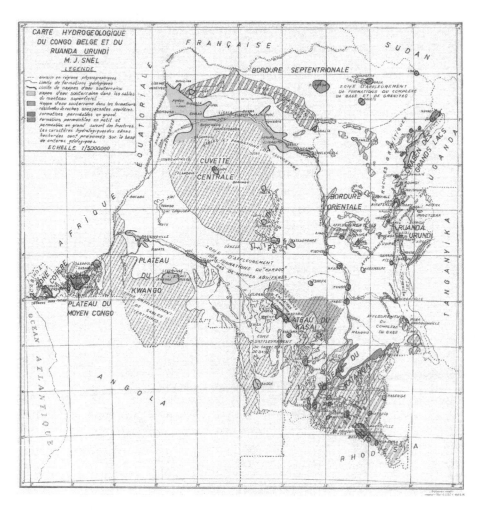

Figure 19.1 1:5 000 000 Carte Hydrogeologique du Congo Belge et du Ruanda Urundi (Snel, 1957).

Table 19.1 Classification system for hydrogeological maps (after Struckmeier and Margat, 1995).

	Level of information		
	---	---	---
	Low *(scarce and heterogeneous data from various sources)*	*Advanced* *(+ systematic investigation programmes, more reliable data)*	*High* *(+ hydrogeological systems analysis and groundwater models)*
Possible use			
Reconnaissance and exploration	General hydrogeological map (aquifer map)	Hydrogeological parameter maps (map sets, atlases)	Regional groundwater systems map (conceptual model representations)
Planning and development	Map of groundwater resource potential	Specialised hydrogeological maps (planning maps)	Graphic representation derived from geographic information systems (maps, sections, perspective diagrams, scenarios)
Management and protection	Map of groundwater vulnerability		

	Possible use	
Parameters of representation	Static-----------------------time-dependence------------------------dynamic	
	Low-----------------------------reliability----------------------------------high	
	Low--------------------------cost per unit area-------------------------high	
	Large------------------------area represented------------------------small	
	Small-----------------------------scale--large	

One of the earliest hydrogeology maps in SADC is that compiled for the Congo, Rwanda and Burundi in 1957 (Snel, 1957) (Figure 19.1). This was followed by a series of national hydrogeological maps, mainly at a scale of 1:1 000 000 and based on the UNESCO standard format, produced in the 1980s–1990s (Table 19.2). Only the maps of Madagascar, Namibia and South Africa were produced using a Geographical Information System (GIS) in a digital format. GIS and Remote Sensing (RS) are basic tools in the discipline of hydrogeology and water-resource development. GIS and RS have been used in a number of countries to determine or delineate areas of good groundwater potential (aquifer productivity) and to quantify recharge. In employing these techniques, a number of thematic fields are considered such as geology, slope, rainfall, geomorphology, DEM, soil types and land use (Bandyopadhyay *et al.*, 2007). Various approaches to the mapping of aquifer productivity are documented in the literature and these approaches were evaluated in the context of the SADC-HGM.

19.2.1.1 Map-compilation methodology

19.2.1.1.1 Base map compilation

The lithostratigraphic base map was compiled using a simplified form of the SADC geology map prepared by the South African Council for Geoscience. This was achieved by combining broad stratigraphic units with a much simplified list of lithologies to produce 12 lithostratigraphic classes (Table 19.3).

Table 19.2 National hydrogeological maps.

Angola	1:1 500 000 Mapa Hidrogeológico De Angola	1990
Botswana	1:1 000 000 Groundwater resources map of the Republic of Botswana	1987
DRC	1:5 000 000 Carte Hydrogeologique du Congo Belge et du Ruanda Urundi	1957
Lesotho	1:300 000 Hydrogeological map of Lesotho	1994
Malawi	1:250 000 Hydrogeological map sheets of:	1987
	Blantyre, Dedza, Karonga, Lilongwe, Mangochi, Mzuzu, Nkhotakota, Nsanje, Nyika	
Madagascar	1:500 000 series of maps	2008
	1:500 000 hydrogeology map	1957
Mauritius	1:50 000 Geological map of Mauritius (2 sheets), hydrogeological map with explanations	1999
Mozambique	1:1 000 000 Carta Hidrogeológico De Moçambique	1987
Namibia	1:1 000 000 Hydrogeological map of Namibia	2001
South Africa	1:500 000 Hydrogeological series of maps	1990s
Swaziland	1:250 000 Hydrogeological map	1992
Tanzania	1:1 500 000 Hydrogeological map	1990
Zambia	1:1 500 000 Hydrogeological map	1990
Zimbabwe	1:500 000 Hydrogeological series of maps	1987

Table 19.3 Lithostratigraphical classes of the SADC HGMs.

Hydro-lithological classes

Unconsolidated sands and gravel – Quaternary, Kalahari, Recent
Clay, clayey loam, mud, silt, marl – Quaternary, Kalahari, Recent
Unconsolidated to consolidated sand, gravel, arenites, locally calcrete,
    bioclastites – Quaternary, Kalahari
Sandstone – Late Precambrian, Karoo, Cretaceous
Shale, mudstone and siltstone – Late Precambrian, Karoo, Cretaceous
Interlayered shales and sandstone – Late Precambrian, Karoo, Cretaceous
Tillite and diamictite – Karoo
Dolomite and limestone – Late Precambrian, Cretaceous
Volcanic rocks, extrusive – Karoo
Intrusive dykes and sills – Precambrian and Karoo
Paragneiss, quartzite, schist, phyllite, amphibolite – Precambrian
Granite, syenite, gabbro, gneiss and migmatites – Precambrian

19.2.1.1.2 *Algorithm for the SADC-HGM*

19.2.1.1.2.1 AQUIFER TYPE AND PRODUCTIVITY

The UNESCO standard legend for groundwater and rocks was adopted (Struckmeier and Margat, 1995), the only difference being the separation of fissured and karst aquifers. This was done because karst aquifers with high-yielding wells are regionally important and are prone to over-exploitation and vulnerable to pollution. The identified lithostratigraphies (Table 19.3) are grouped into four main aquifer types where they can be related to a degree of productivity within a matrix to define eight mappable units (Table 19.4).

19.2.1.1.2.2 ASSIGNING AQUIFER PRODUCTIVITY

The process used to assign productivity includes:

- evaluation of borehole data to produce typical aquifer hydraulic properties;
- assessment of recharge distribution;.
- expert assessment of data and outputs.

The algorithm for the SADC-HGM took into account data scarcity and patchy distribution. Some countries were unable to provide either the required data or information. Assessment by experienced national and regional hydrogeologists was used to infill data gaps during the development of the hydrogeological map under the supervision of the project steering committee.

The lithostratigraphic units identified (Table 19.3) were grouped into permeable and low-permeability formations by applying borehole-derived aquifer parameters and information from area-specific hydrogeological studies including National Water Master Plans. Using these data, the permeable formations are grouped into porous (gravel, alluvium, sand, etc.), fissured (sandstone, basalt, etc.) and karst (limestone, dolomite, gypsum, etc.) classes. Aquifer types shown on the completed map are:

- unconsolidated intergranular aquifers;
- fissured aquifers;
- karst aquifers;
- layered aquifers;
- low-permeability formations.

Eight aquifer productivity types were identified (Table 19.4) using the methodology proposed by Struckmeier and Margat (1995). The classification combines

Table 19.4 Aquifer-productivity domains.

Class aquifer type	Productivity			
	1. High productivity	*2. Moderate productivity*	*3. General low productivity but locally moderate productivity*	*4. Generally low productivity*
A. Unconsolidated intergranular aquifers	A1	A2	X	X
B. Fissured aquifers	B1	B2	X	X
C. Karst aquifers	C1	C2	X	X
D. Low-permeability formations	X	X	D1	D2

Denotes an extensive aquifer overlain by cover

information on aquifer productivity (lateral extent) and the type of groundwater flow regime (intergranular or fissured).

Assignment of aquifer productivity to the different lithostratigraphic units (Figure 19.2) is based on flow properties (transmissivity) and sustainability (local recharge). As an example, moderate recharge conditions combined with a highly transmissive unconsolidated intergranular aquifer were assigned as A1, a productive aquifer. The long-term aquifer productivity of each aquifer domain depends upon the inherent lithological properties (i.e. conductive material properties), hydraulic parameters and groundwater through flow (i.e. groundwater recharge to the formation). These aquifer-productivity domains can be classified by these basic parameters. Aquifer productivity domains in the upper right corner of the matrix are more productive than those of the lower left corner. In addition, aquifer-productivity domains to the left require boreholes over a larger area than the domains to the right.

The combination of transmissivity and recharge may be broadly used to indicate aquifer productivity. For instance, high transmissivity and low recharge may imply:

- a high short-term aquifer yield;
- a low long-term productivity;
- high potential for borehole failure under drought conditions.

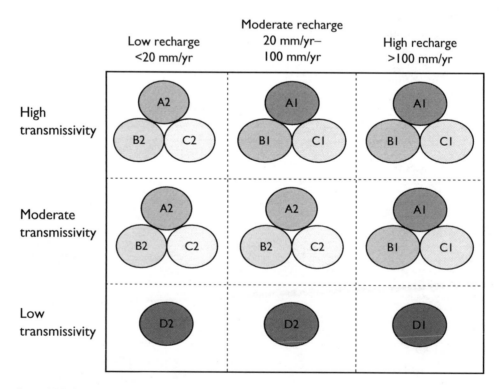

Figure 19.2 Scheme adopted for assigning aquifer-productivity domains on the SADC hydrogeological map. Refer to Table 19.5 for the aquifer-productivity classes.

In contrast, low transmissivity and a high recharge may imply:

- long-term sustainable low yields from many boreholes;
- high potential for sustainable low-yield boreholes during drought periods.

The water balance must be assured in the long term, which implies that the long-term aquifer productivity is limited by the magnitude of recharge to the aquifer. The productivity map should only be seen as a starting point from which countries should be able to update the information whenever new field data become available and thus highlights the importance of monitoring water levels to assess the impact of long-term abstraction.

The assignment of aquifer productivity is a lengthy process that takes into account all available geological and hydrogeological data for each area (a polygon or a set of polygons). In effect, the lithostratigraphic type polygons were overlain, within a GIS or manually, with relevant reference layers (mainly the scanned national hydrogeological maps), and an essentially 'manual', expertise-based decision was made for each area to compile the SADC-HGM (Figure 19.3).

19.2.1.1.2.3   GROUNDWATER RECHARGE

Information on groundwater recharge is a fundamental component of any hydrogeological map. The recharge data layer is necessary for the assignment of aquifer productivity, or is portrayed as an inset map. The derivation of a groundwater-recharge layer was investigated based on an algorithm to be applied at a regional SADC scale. However, recharge is a complex process governed by a number of controlling factors (Figure 19.4) that are highly variable in space and time (Bredenkamp *et al.*, 1995). As data for these variables are not readily available on a region-wide scale within the SADC, it was not possible to develop a recharge algorithm for the SADC-HGM within the scope of the project.

19.2.1.1.2.4   TRANSBOUNDARY AQUIFER SYSTEMS

Groundwater systems, including transboundary aquifer systems (TBAs), cannot be delineated by lithological domains alone. Detailed information is also needed on factors such as topography, soils, depths of overburden, groundwater flow, extent of fracture systems, and identification of recharge and discharge areas. Comprehensive collections of such information and data are rarely available especially for a region as diverse as the SADC area. As a consequence, transboundary aquifer map layers on global and regional scales are commonly broadly marked as circles and ellipses, the most recent one being the ISARM (2012) Atlas of Transboundary Aquifers where 17 transboundary aquifers systems between SADC countries are listed (IGRAC, 2012). These are also shown on the World Hydrogeological Map, WHYMAP. In 2005 the International Groundwater Resources Assessment Centre (IGRAC) presented a map layer in an attempt to delineate transboundary aquifers based on lithological transitions and catchment boundaries (Vasak and Kukuric, 2006).

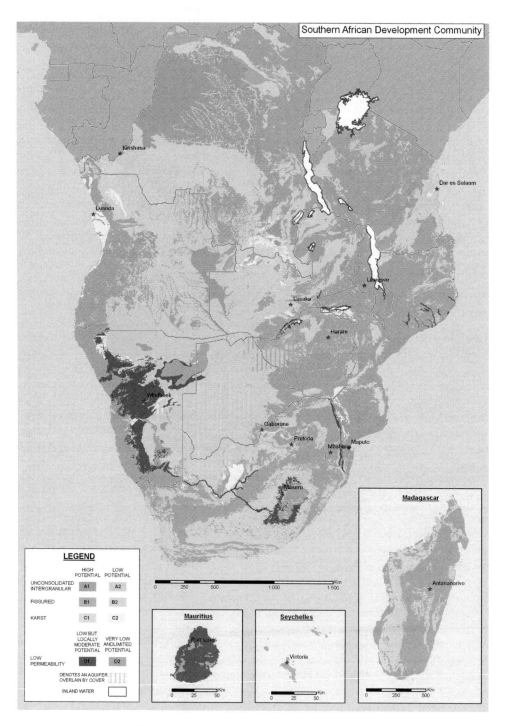

Figure 19.3 SADC-HGM (here represented at scale 1:30 000 000).

Unsaturated zone

- soil type and thickness
- slope
- moisture
- geology
- Fracturing and tectonics

Rainfall

- volume
- intensity
- frequency
- areal distribution

Evapotranspiration

- temperature and wind
- humidity
- soil type and
  soil moisture
- vegetation (type
  and density)

RECHARGE

Figure 19.4 Main elements and controlling factors of groundwater recharge.

In the compilation of the SADC-HGM the criteria used for delineating transboundary aquifers were:

- shared by more than one SADC country;
- continuous transmissive aquifers;
- sub-basin river boundaries;
- SADC hydro-lithological boundaries.

Table 19.5 Transboundary aquifers.

Name	Code	States
Karoo Sandstone Aquifer	6	Tanzania, Mozambique
Tuli Karoo Sub-basin	15	Botswana, South Africa, Zimbabwe
Ramotswa Dolomite Basin	14	Botswana, South Africa
Cuvelai and Etosha Basin	20	Angola, Namibia
Coastal Sedimentary Basin 1	3	Tanzania, Mozambique
Shire Valley Aquifer	12	Malawi, Mozambique
Congo Intra-cratonic Basin	5	D R Congo, Angola
Coastal Sedimentary Basin 2	4	D R Congo, Angola
Coastal Sedimentary Basin 6	21	Mozambique, South Africa
Medium Zambezi Aquifer	11	Zambia and Zimbabwe
Dolomitic	22	D R Congo, Angola
Sands and gravel aquifer	23	Malawi, Zambia
Kalahari/Karoo Basin	13	Botswana, Namibia, South Africa
Eastern Kalahari/Karoo basin	24	Botswana and Zimbabwe

The identified transboundary aquifers are given in Table 19.5 and Figure 19.5. The extent of these aquifer systems needs to be verified thorough research and extensive field investigations.

Davies *et al.* (2012) classified the TBAs identified in the SADC-HGM on the basis of five basic data sets and as a result proposed three classes of TBAs (Table 19.6).

The assessment concluded as follows (Davies *et al.*, 2012):

Troublesome TBAs (likely to be the cause of friction between neighbouring states):

TBA 15, Tuli Karoo Basin Aquifer

TBA 24, Eastern Kalahari Karoo Basin Aquifer.

Less troublesome TBAs:

TBA 13, South West Kalahari/Karoo Basin Aquifer

TBA 14, Zeerust-Ramotswa-Lobatse Dolomite Basin Aquifer

TBA 20, Cuvelai Delta and Ethosha Pan Alluvial and Kalahari Sediments.

Moderately troublesome TBAs:

TBA 3, Ruvuma Delta Coastal Sedimentary Basin Aquifer

TBA 4, Congo Delta Coastal Sedimentary Basin Aquifer

TBA 6, Tunduru/Maniamba Basin Karoo Sandstone Aquifer

TBA 11, Middle Zambesi Rift Upper Karoo Aquifer

TBA 12, Shire Valley Alluvial Aquifer

TBA 23, Sands and gravels of weathered Precambrian Basement Complex.

Inactive TBAs (unlikely to become the cause of friction between neighbouring states):

TBA 5, Congo/Zambesi basins Benguela Ridge Watershed Aquifer

TBA 21, Coastal Tertiary to Recent Sedimentary Basin Aquifer

TBA 22, Lower Congo Precambrian Dolomite Aquifer.

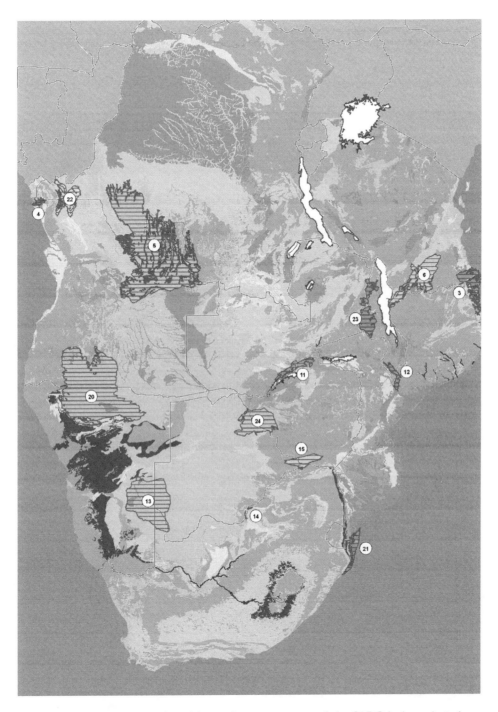

Figure 19.5 Transboundary aquifers delineated as an outcome of the SADC hydrogeological map compilation. See Table 19.5 for codes.

Table 19.6 TBA classification after Davies *et al.* (2012).

Basic data sets

1. Groundwater flow and vulnerability
   a. Natural flow
   b. Induced flow
   c. Aquifer vulnerability
2. Groundwater knowledge and understanding
   a. Groundwater Quantity
   b. Groundwater Quality
   c. Vulnerability
3. Governance capability (Table 19.5)
   a. Management
   b. Knowledge
   c. Monitoring
4. Socio-economic/water demand capability (Table 19.6)
   a. Demography
   b. Land Use
   c. industry
5. Environmental capability (Table 19.7)
   a. Hydrology
   b. Sustainability
   c. Climate

Classes

- **active** in which some form of international collaboration in monitoring, management and apportionment are needed now in order to avoid confrontation in the future should demography, land use or climate be changed.
- **moderately active** in which there is potential for transboundary degradation of some form or another, although it does not currently require international collaboration, i.e. the potential for degradation is so small it will not impact communities either side of the border.
- **Inert or non-active** in which there is no apparent potential for cross border degradation or any impact of any kind.

19.2.1.1.3   Final hydrogeological map and atlas

The map compilation and associated cartography are presented in a GIS environment that allows integration of the draft interactive web-based SADC Hydrogeological Map and Atlas by users on-line. The map is published through a map server, within which additional background data layers including a shaded relief background, topography, surface waters, roads, major towns, layer and satellite imagery are available.

19.3   CONCLUSION

19.3.1   Use and application of the map

The continued use and application of the SADC-HGM by member states will ensure the sustainability and the periodic upgrading of the hydrogeological map as more data of the correct quality become available. The map should be used to:

- guide policy-making and influence political decision-making on natural resource issues;
- provide support in transboundary groundwater planning and management;
- create awareness related to groundwater issues;
- assist in regional development planning;
- serve as a platform for those countries that want to update their national hydro-geological maps.

19.3.2   Updating the map

The SADC-HGM is a general hydrogeological map, which provides information on the extent and geometry of regional aquifer systems. The map is intended to serve as a base map for hydrogeologists and water-resource planners, whilst at the same time informing non-professionals of groundwater conditions. The map is a visual representation of groundwater conditions in the SADC and serves as a starting point for the design of more detailed regional groundwater investigations by exposing data and knowledge gaps. It is important to note that the SADC-HGM is published as an interactive web-based map and not as a printed map. The SADC-HGM will require periodic updating. Key to updating the map is the improvement of groundwater data sets and information systems in the various countries. Any future update of the map requires a bottom-up approach to work with countries to ensure that representative data sets are obtained from the various geological domains.

19.3.3   Lessons learnt

The main lessons highlighted by this project include:

- a stable, long-term and secure environment for the storage, updating and continued dissemination of the map needs to be obtained. Issues to consider are:
  - documentation – staff members come and go, so process documentation should exist to allow new staff to maintain and update the map;
  - backup of all digital files.
- a high-quality set of borehole data is central to hydrogeological mapping. While the critical process of assigning productivity ratings to discrete lithological areas (map polygons) should ideally use borehole data (specific capacity, rest water level, borehole depth, more direct indications of productivity if available), it is possible to incorporate expert knowledge and inspection of existing maps as well;
- during the course of the project it was noted that many of the member states do not maintain a complete borehole and hydrogeological database, for a number of reasons, with cost and institutional fragmentation being important ones. The importance of such a central, national-level database cannot be over-estimated, not only for the production of maps, but generally for management of groundwater resources;
- publishing a digital map to the web involves many new design decisions, hosting hardware and software decisions, and issues of bandwidth, web access by users,

permissions and web-site management. It should be noted that the effort to publish maps on the web requires considerable technical expertise at start-up time, considerable bandwidth from the web-server and by likely users, and adequate funding for sustaining the initiative. An alternative that should be considered is the copying of the GIS-based dataset to a CD or DVD.

19.3.4   Opportunities for the future

- inform and advertise the product to stakeholders at conferences and forums. In this regard a printed map of the SADC-HGM will be advantageous and a project should be initiated to achieve this outcome;
- implement the capacity-building strategy that was developed as part of the project;
- ensure that the hydrogeological mapping programme is included in the future update of the Regional Strategic Action Plan;
- make the information and data available to programmes that are supporting the SADC objectives;
- improve field procedures to collect groundwater data and information;
- develop better understanding of the hydrogeological properties of the various geological domains;
- the various techniques employed in determining aquifer productivity are technical and require professionals with both extensive hydrogeological expertise as well as some relatively good GIS skills. Training of staff from the member countries in such techniques is thus critical in the context of sustainability;
- aquifer recharge is a complex field that requires high-level skills which are lacking in most of the SADC member states. Staff members from member states therefore need training in aquifer-recharge determination and processing;
- member states should be encouraged and supported to embark on national hydrogeological mapping programmes; this would significantly benefit the objectives of the SADC-HGM as well as future updates of the SADC Hydrogeological Map itself.

ACKNOWLEDGMENTS

The SADC Hydrogeological Map has been in planning since the early 1990s. Numerous persons have been involved in the conceptualisation of the map, since those days, and their contributions to the project are gratefully acknowledged. We express our gratitude to the project steering committee members and country participants for their input during the course of the project. The project team was hosted by the Department of Geological Survey in Lobatse, Botswana. The SADC Project Manager, Mr O Katai, is thanked for his leadership throughout the project. Mr Jeffrey Davies (BGS) and Dr Willi Struckmeier (BGR) are thanked for the valuable advice during the course of project implementation. The SADC Infrastructure and Services Directorate – Water

Division is also thanked for their contribution to the project. We are grateful to the European Union and GIZ for their financial support to the project.

REFERENCES

Bandyopadhyay, S., Srivastava, S.K., Madan, K.J., Hedge, V.S. & Jayaraman, V. (2007) Harnessing earth observation (EO) capabilities in hydrogeology: an Indian perspective. *Hydrogeology* 15, 155–158.

Bredenkamp, D.B., Botha, L.J., Van Tonder, G.J. & Van Rensburg, H.J. (1995) *Manual on Quantitative Estimation of Groundwater Recharge and Aquifer Storativity*. Pretoria, South Africa, Water Research Commission.

Davies, J., Robins, N.S., Farr, J., Sorensen, J., Beetlestone, P. & Cobbing, J.E. (2012) Identifying transboundary aquifers in need of international resource management in the SADC region of southern Africa. *Hydrogeology Journal*.

Hartzer, F. (2009) *Geological Map of the Southern African Development Community (SADC) Countries, 1:2 500 000 scale*. Pretoria, South Africa, Council for Geoscience.

IGRAC (2012) *Transboundary Aquifers of the World*. Delft, The Netherlands, IGRAC.

SADC (2005) *Regional Strategic Plan on Integrated Water Resources Development and Management. Annotated Strategic Action Plan 2005–2010*. Gaborone, Botswana, South African Development Community.

Snel, M.J. (1957) *1:5 000 000 Carte Hydrogeologique du Congo Belge et du Ruanda Urundi*. Congo, Service Geologique Leopoldville.

Struckmeier, W.F. & Margat, J. (1995) *Hydrogeological Maps A Guide and Standard Legend* (Vol. 17). Verlag Heinz Heise, Hannover, Germany, International Association of Hydrogeologists.

Vasak, S. & Kukuric, N. (2006) *Groundwater Resources and Transboundary Aquifers of Southern Africa*. Utrecht, Netherlands, IGRAC.

Author index

Subject index

Series IAH-selected papers